About Island Press

Since 1984, the nonprofit organization Island Press has been stimulating, shaping, and communicating ideas that are essential for solving environmental problems worldwide. With more than 1,000 titles in print and some 30 new releases each year, we are the nation's leading publisher on environmental issues. We identify innovative thinkers and emerging trends in the environmental field. We work with world-renowned experts and authors to develop cross-disciplinary solutions to environmental challenges.

Island Press designs and executes educational campaigns, in conjunction with our authors, to communicate their critical messages in print, in person, and online using the latest technologies, innovative programs, and the media. Our goal is to reach targeted audiences—scientists, policy makers, environmental advocates, urban planners, the media, and concerned citizens—with information that can be used to create the framework for long-term ecological health and human well-being.

Island Press gratefully acknowledges major support from The Bobolink Foundation, Caldera Foundation, The Curtis and Edith Munson Foundation, The Forrest C. and Frances H. Lattner Foundation, The JPB Foundation, The Kresge Foundation, The Summit Charitable Foundation, Inc., and many other generous organizations and individuals.

The opinions expressed in this book are those of the author(s) and do not necessarily reflect the views of our supporters.

Getting to the Heart of Science Communication

Getting to the Heart of Science Communication

A Guide to Effective Engagement

Faith Kearns

 **ISLAND**PRESS | Washington | Covelo

© 2021 Faith Kearns

All rights reserved under International and Pan-American Copyright Conventions. No part of this book may be reproduced in any form or by any means without permission in writing from the publisher: Island Press, 2000 M Street, NW, Suite 480b, Washington, DC 20036.

Library of Congress Control Number: 2020944650

All Island Press books are printed on environmentally responsible materials.

Manufactured in the United States of America
10 9 8 7 6 5 4 3 2 1

Keywords: Island Press, science communication, engagement, public scholarship, relational, listening, conflict, power, accountability, trauma, equity, inclusion, self-care, ethics, standpoint, training, practice

For everyone doing the work

Contents

Foreword xi

Preface xv

Acknowledgments xix

Introduction   1

Part I. The Evolution of Science Communication

Chapter 1. Science Communication from the Ground Up   11

Chapter 2. Science Communication Careers Today   29

Chapter 3. Navigating Facts and Feelings in Science Communication   55

Part II. The Tools of Science Communication

Chapter 4. Relating   79

Chapter 5. Listening   103

Chapter 6. Working with Conflict   133

Chapter 7. Understanding Trauma   151

Part III. The Future of Science Communication

Chapter 8. Equitable, Inclusive, and Just Science Communication   173

Chapter 9. Self-Care and Collective Care   203

Chapter 10. What More Is Possible?   227

Notes   235

About the Author   255

Foreword

We are in a moment of great upheaval.

As a scientist and as a human, I've been watching with anguish for years as the climate emergency worsens. This has happened despite a strong scientific consensus that human actions are the main driver of dangerous environmental change.

Clearly, it's not enough just to be correct and repeat the facts in the middle of an existential crisis such as this. This crisis is getting worse not because of a deficit of information, after all, but because of a deficit of action. Eventually, I realized I couldn't separate my desire to understand and describe the world around me in a factually accurate way from the special demands this moment placed on me as a science communicator. I had to learn how to tell the truth in a new way.

My personal transformation from scientist to science communicator started by asking myself these questions:

Whom do I do science for?

How does my work create or diminish injustice?

How can I help my fellow scientists bridge into revolutionary change?

In this book, Dr. Faith Kearns has provided a guide for how to start this conversation for yourself and, in so doing, make your science communication more effective.

This is difficult, demanding work, and I'm only at the beginning of it. Experts like Dr. Kearns are at the cutting edge of understanding what this moment requires of science communicators embedded in a new science communication landscape, where people are grappling with emotions and conflict in the midst of huge uncertainties.

To get the broader science community to come along, it will be necessary to have uncomfortable conversations. That's okay. In fact, it's the only way change like this can happen. As Dr. Kearns writes, "What the debate over science advocacy has regularly glossed over is that there are at least as many ethical concerns with standing on the sidelines as there are with engaging."

The risks of speaking up, of course, are not equal. Women and people of color bear the brunt of the cost of attempting to reframe stale and ineffective communication methods to center justice. Not by coincidence, women and people of color are also bearing the brunt of the impacts of the climate emergency, because of a pervasive culture of white supremacy and toxic masculinity that has infected almost every aspect of how science is performed and how stories about science are told. The good news is, none of this is inevitable. Change is possible, and it must happen in order to build a better world we are all worthy of.

Justice-centered communication is dangerous work for scientists who want to serve society. Science communication practitioners are evermore precariously employed while also navigating politics and unbalanced power dynamics in their institutions, families, and communities.

The kinds of skills justice-centered science communicators need are not anything like the ones being taught. We need to be able to relate, listen, work with conflict, and understand the role of trauma in the communities we serve. That even talking about science communication in this way upsets people who are hindering necessary progress means that this is clearly powerful work.

But this is the work worth doing. We need everyone.

Scientists need to participate in creating a new world, and to do so they will need to shed some of their traditional reliance on authority and objectivity to truly link arms with others engaged in larger struggles. This book provides some initial guidance on doing so.

We need courage in our time of greatest need. We need to work in solidarity with one another, and this book will help all of us along our intertwined journeys.

Eric Holthaus

Preface

This is a book written from the point of view of a scientist who has also been a workaday science communication practitioner for over twenty-five years. It is not a research book, not an academic tome, not a literature review but instead an attempt to capture the grounded, lived experiences of practitioners. Despite a long career, I rarely find my experience and those of many of my peers reflected in the science communication research literature or in advice written primarily by journalists. The material and narratives the book contains come from those actively working on the ground, facing the daily joys and dilemmas of science communication on what can be emotional and contentious issues, while they are also directly accountable to communities.

If you consider yourself a science communicator or someone interested in how scientists and other technical experts communicate research, my hope is that what I, along with the many people graciously willing to be included here, add to the conversation will be of use. You may be a traditional or nontraditional academic, work at a government agency or consulting firm or nonprofit, or do many other kinds of work. If you are interested in the topic, this book is for you and will provide an alternative to the conventional science communication conversation of the last few decades, which focused mainly on helping scientists connect with journalists and decision-makers.

To ease the reading, I would like to clarify two pieces of terminology:

The first is that I use the terms *science communication* and *engagement*—as well as the occasional variants *outreach*, *extension*, and *public scholarship*—somewhat interchangeably. This language is purposeful. I have in my own career tried to separate communication and engagement, only to circle back to feeling that the exercise has not been useful. At its best, the science communication tent is broad, and I have no interest in being exclusionary.

In addition, I use the term *practitioner* throughout the book to mean "a person who practices science communication," whether full- or part-time, in any context. The terms *practice* and *practitioners* are used to distinguish science communication practice from science communication research and researchers. Where the work is done—whether at a university, in an agency, or with a nonprofit—is not a defining feature of a practitioner. In fields like law and medicine, the term *practitioner* is ubiquitous. It is not yet widely used in many science fields; however, as the sciences move closer to fields with distinct research and practice roles, taking up the term will be increasingly useful.

I want to acknowledge that writing about anything in this first part of the twenty-first century is a challenge. We are on shaky ground and stumbling around for truths that might be irrelevant tomorrow or next week, much less next year—particularly as the coronavirus has emerged. As I write this, many of the terms I use are already falling by the wayside and being replaced with others. For example, in a single day, I can read pieces vehemently arguing that we need to move beyond diversity to talk about inclusivity, only to read other pieces critiquing the concept of "being included" in a culture that is unjust. These are necessary evolutions, but they do make describing things complicated.

It is also important to note my positionality as the author of this text; that is, some of the many factors that influence my perspective, which is necessarily limited. I am a cisgendered white woman in my midforties. I earned a doctoral degree from the University of California, Berkeley, in environmental science and policy. I now hold a term-renewable, non-tenure-track academic position after moving between government and the nonprofit sector. I am divorced and in a long-term partnership. I was raised in a working-class family in rural Arizona. I am the first person in my family with a doctorate, though a large part of my childhood was spent living on a college campus as my parents undertook their bachelor's degrees. I graduated prerecession, yet between student loans and other obligations, I remain a renter. With this text, I am committing to learning in public, to knowing that I will always have more to understand, and to updating my views as my understanding grows. My goal is to further a conversation rather than have the last word.

The narratives shared here are ones that I can attest to personally or those that have been shared with me but are of course only a small fraction of what could be told from well beyond my small part of the world, which is part of a much larger global community of science communicators. There is a tendency in the sciences to universalize and generalize, and sometimes that makes sense, but for me it doesn't. Part of what I'm trying to describe is the complexity in science communication, particularly when it is highly localized and built on relationships that necessitate navigating power and accountability. The best way I can do that is to convey the complexities of specific people, locations, and issues with careful consideration given to their specificities. I trust you, the reader, to draw on what is relevant to your own context and fill in the blanks with your own experiences and questions.

A single book cannot be all things to all people, but my hope is that I can fairly treat the argument that emotion, conflict, and power struggles are already present in science communication and engagement work and that ignoring them is counterproductive. That means I am navigating different reader experiences simultaneously. I believe that conflict, for instance, is an important process in science communication, but the experience of conflict is not the same for everyone. That may mean that people who love conflict and receive the benefit of the doubt when angry and in conflict need to step back from it, while those of us who have a hard time being angry in public might need to learn to step into righteous anger when it is called for.

Finally, I am providing less a guide than a map—to further conversations rather than shut them down, to introduce some answers but raise more questions. I actively live with these issues every day myself and look forward to working to figure things out, together.

Acknowledgments

This book is in many ways a work of community, and I am deeply grateful to everyone who spoke with me, formally and informally, over the past dozen or so years and contributed, directly and indirectly, to its contents. Many of those people are featured in its pages. This book would not have been possible without every single person who consented to share a part of their heart, mind, and experiences with me, as well as the work done by the scholars and practitioners whom I have cited here.

I appreciate the support of Doug Parker at the California Institute for Water Resources for providing me with some time and space for this effort. I am grateful beyond measure to everyone who took the time to read my draft or portions thereof and to give generous feedback. My mother, Deborah Kearns, is my first reader and most enthusiastic supporter, and I am endlessly thankful to her for everything. My partner, Art Medlar, is my second and most critical reader, and for that I am also thankful.

Writers John Fleck, Cynthia Barnett, and Eric Holthaus gave me excellent advice at a couple of critical junctures, including getting started. Erin Johnson is a helpful editor, and I am thankful to her for agreeing that the original idea had some promise and to the entire team at Island Press that made this book possible. Rose Hayden-Smith was an irreplaceable cheerleader in helping me get the writing off the ground. Lucy Andrews, Dezaraye Bagalayos, Sarah Boone, Clare Gupta, Kripa

Jagannathan, Sara Kahanamoku, Yanna Lambrindou, Jennifer Lentfer, Nadine Lymn, Marilyn Myers, Sarah Myhre, and Sarah Reed provided valuable feedback, as did everyone I interviewed who read and commented on the sections, if not whole chapters, that included them.

It's always hard to see where something so close to you begins, but as good a place as any is with the people who taught me—with love and courage over many, many years—what it means to work in relationship, even when the work is difficult. Thank you to Leslie Brown, Mary Barber, Nadine Lymn, Gail Silverstein, and Dorothy Stump, always. I have also been blessed throughout this process with the loving memory of my grandmother Lenore Foster.

Finally, I have deep gratitude for every place that held me during this process. This includes the kitchen table of my tiny studio in Berkeley, land of the Ohlone, where I did most of the work; along with the oak woodlands to my north, where I spent crucial time; and the deserts south and east, continual reminders of what sustains.

Introduction

As I walked the short distance from the plane to the terminal, the humidity permeating the collapsible walls of the jet bridge signaled just how far I had come from the dry high-mountain desert of Arizona where I was raised. I arrived in Washington, DC, not for the first time but for what would turn out to be one of several times I moved there for a job.

Within a few short weeks, I sat at the National Press Club in front of a panel that was discussing the release of the second report from the Intergovernmental Panel on Climate Change. It was 1996, and while I had certainly learned about greenhouse gases as an undergraduate environmental science major, little did I know how the information being discussed that day would shape much of my professional life.

As far as I can recall, I did not walk away feeling worried—the presentation by scientific experts conveyed calm competence, as intended. At the time, climate change was commonly grouped into the same bucket as acid rain and ozone depletion, both problems that had been successfully addressed through a series of policy agreements with what is now remembered as generally widespread and bipartisan support. My young self was assured that climate change would be dealt with in a similar fashion—with scientific authority, regulation, and cooperation.

Fast-forward a couple of decades, and I find myself living in a world that often feels as if it were burning. Then there are the days, weeks, months even when it actually is. As a scientist and communications practitioner working at the intersection of water, fire, and climate change in the western United States, I am often moving from one disaster to another, both literally and figuratively. And while I feel camaraderie with the relatively small group of other disaster specialists, the COVID-19 pandemic has delivered the kind of urgency and uncertainty common to disasters to a much broader swath of people around the globe.

I look back at how easy, albeit naive, it was to assume that the scientific information being generated about climate change would be acted upon swiftly, rationally, collaboratively. I didn't have reason or experience enough to think otherwise. I now have very different tales to tell and lessons to share that apply to topics well beyond climate change, stretching into the many emotional and contentious matters that scientists and science communicators work on today.

Learning Science and Communication

When I began my scientific training, I could not have imagined that my delight in learning about the ancient life embedded in the diverse formations of the Grand Canyon would lead me to where I am today. As an undergraduate at the small public university near my hometown, I was fortunate enough to be interested in one of the subject areas that the dramatic landscapes on the edge of the Colorado Plateau were best suited for: geology. That meant I got to spend several afternoons a week immersed in deep time during field labs at sites ranging from the many dormant volcanoes to vast synclines that were within a relatively short distance of our campus.

I had also been able to get a work-study position in the Athletics department of the university, a clear improvement over my high school jobs. I started as a student administrative assistant my first year, and by my second was working first in fundraising and events and then in marketing and communications. In this professional, virtually full-time, and very fun job, I learned about communications by doing communications. For four years, I wrote press releases and recorded public service announcements; designed flyers, T-shirts, and magnets; and invented halftime events.

As a science major, I was asked by peers and professors why I didn't try to get a job more befitting my scientific interests. And at work, I was asked what I was doing majoring in science when communications was obviously my passion. I never had a good answer; the term *science communication* was not yet being used, and there was no training to take or online forum in which to meet the small, loose group of people doing related work.

However, when one of my professors handed me an announcement for a modestly paying public affairs internship with the Ecological Society of America (ESA), a path started to form. I applied for the internship and, much to my surprise, was offered the position. As only my twenty-two-year-old self was capable of doing, and armed with a vastly overextended credit card and the help of family friends who allowed me to live in their basement for a few months until I could afford my own apartment, I picked up and moved across the country, and it suddenly became evident how my interests fit together.

I spent the summer skimming abstracts for the society's annual meeting, writing press releases on several of the most interesting studies that would be presented there, and attempting to get journalists to write about these studies. It was a thrill to have publications ranging from *Discover* magazine to the *New York Times* cover research I had written about.

I also designed and wrote research briefs on how long-term ecological research had led to breakthroughs in our understanding of both Lyme disease and hantavirus. I was exhilarated to be able to combine what had seemed like widely disparate interests in science and communication into something that was starting to look as though it had a future.

I was, thankfully, able to parlay this three-month internship into a permanent position with ESA, working with what was known as the

Sustainable Biosphere Initiative. That office arose from ideas in a paper led by Jane Lubchenco, a professor at Oregon State University who would go on to cofound the science communication training organization COMPASS in 1999.

I continued my communication work in that position, organizing and summarizing research symposia, creating the design for a new publication series, and enthusiastically writing about whatever I could. It became apparent to me that I liked this work but that in order to do it at the level that I aspired to and to be taken seriously while doing so, I was going to need to go to graduate school.

I embarked on my doctoral program at the University of California, Berkeley, pursuing a traditional research degree. However, I struggled with the specificity needed to complete my work on urban freshwater ecology. During my second year, I was fortunate to begin working with a new cooperative extension specialist in my department.

The learning curve was steep, as I had to quickly learn about cooperative extension, but I sensed I would be able to do the kind of communications work that interested me. In that context, my expertise expanded into participatory science and mapping, which allowed me to keep doing design work, primarily for the web. After an interesting year spent as an American Association for the Advancement of Science (AAAS) Science and Technology Policy Fellow at the US State Department postgraduate school, I returned to UC Berkeley in a non-tenure-track academic position to continue my interests in web mapping with a newly formed wildfire research and outreach center.

There, my ideas about science communication shifted significantly, in ways that took many years to understand. As just one example, at a community fire safety day in Northern California in 2008, I had a conversation with a man upset about the way my colleagues and I talked about fire.[1] In retrospect, I understand that he told me exactly what I needed to hear, which was, in essence, that we approached the matter with an alarming level of intellectual distance in a place where people were living with the reality of fresh wildfire.

Just months before, a large cluster of wildfires had burned through this small, close-knit community, forcing many residents to evacuate their homes. In a story that sounds familiar to many today, they had been worried that in their absence, the properties would go unprotected because firefighting resources were spread thin. They had reason to worry: there were fires burning all over the state that year, and firefighting resources were indeed stretched particularly thin.

As we presented our research on how houses could be built to withstand fire, as well as the controversial idea of being trained to stay with your house to keep it from burning, the mood in the room became increasingly somber. From a science standpoint, it was pretty straightforward stuff: We had a lot of great research and some nice tools, including interactive checklists and hazard maps to make the work accessible. We had been working with local firefighters and communities, state agencies, and legislative staff to figure out how to implement our work. We were doing everything we were "supposed" to do as community-engaged scientists.

But here we were, talking to people who had just lived through traumatic events that intersected with our research on evacuation, and this man was kind enough to share with me how hearing about our work in the way that we'd presented it had been retraumatizing for him. He was telling me something important—his emotions came through plainly enough for me to hear that. I was able to hear him in part because, as many people do, I had faced a crisis in the middle of my graduate training that, thankfully, led me to a transformational therapeutic relationship.

With skills I had gained in therapy and through various contemplative practices, I was able to really hear this man. Listening, truly listening, to him was the first time I became acutely aware that the intellectual rigors of my scientific training had not prepared me for the strong bouts of emotion that come with sharing research that has immediate meaning in people's lives. Speaking with him inspired years of work to understand how scientists, and the communities they are a part of, can relate in ways that are responsible and generative.

After those potent years spent working on fire in California, and in the midst of the recession, I returned to DC to pick up where my AAAS fellowship had left off, working on contentious fisheries and related marine affairs in a new position working between the advocacy and science programs at the Pew Charitable Trusts. In my current role with the California Institute for Water Resources in the University of California's Division of Agriculture and Natural Resources, I am back to working on water, fire, and climate change.

In each of these boundary-spanning positions, I dealt with a wide variety of emotional and controversial topics in many different settings, and my ideas about communication have been tested again and again. I saw gaps in the way that I communicated that only grew with time, primarily because I had taken on the never-spoken but omnipresent message that feelings aren't allowed in science. I now know that shortfall was in part the effect of what is called the hidden curriculum—a rich concept encompassing all the things we implicitly learn in the training process that deeply impact our worldviews and ultimately can shape our entire careers.

However, it turns out that people (myself included) have a lot of feelings about fire and fish and water and many other scientific subjects—today, a cultural reference point might be the debate over the effectiveness of wearing masks during the coronavirus pandemic—that directly influence the effectiveness and experience of science communication efforts. Sometimes it's not only feelings but also literal life-and-death matters that can cause a traditional researcher to question pretty much everything they've learned, as I did.

I was doing science communication before it got its name, and in my twenty-five-plus years working in science and communication and science communication, I have seen the fields change vastly in some ways while stagnating in others. This book is a distillation of what I have learned and want to share.

The Potential for Science Communication and Engagement

What I will argue in this book is that although traditional science communication methods are valuable, the issues playing out in the twenty-first century are inviting new ways of thinking about and doing science communication and engagement. That is to say, the early focus on information dissemination and science communication as a one-way performance is ceding ground to more engaged and relational communication techniques that also address power and accountability. This examination of the standard assumptions of the field is necessary.

The people doing science and science communication work are also changing. No longer the primary domain of research faculty, the field of science communication is growing into a profession. This change has implications for both science communication and the scientific enterprise, as well as for the communities in which we all live and work.

In the first section of this book, I outline the changing nature of science communication and engagement, including the impact of changing science careers, as well as the emotion, conflict, and power struggles that are present for many scientists and science communicators.

In the second section, I first describe an approach to rethinking science communication and engagement as a true *relationship* challenge and what that means for training and practice. In the three chapters that follow, I assume the reader sees relationship-centered communication as valuable and wants to understand some of the key tenets of that work: listening; working with conflict and power; and understanding the roles of trauma, loss, grief, and healing.

The final section focuses on issues that arise more broadly in science communication practice. I address the role of who is actually doing the work in science communication practice, which has been generally overlooked until recently, and I discuss how to balance working on deeply systemic issues while also taking care of yourself and the interplay between the two.

I hope that readers will gain a new perspective on science communication and engagement that, while in many ways has always been present, offers a way forward on some of the thorniest matters of the day. This book is not meant to be a replacement for any other work but instead offers an additional "view from the ground."

PART I

The Evolution of Science Communication

CHAPTER 1

Science Communication from the Ground Up

THREE NEW WILDFIRES HAVE STARTED in California this week, adding to news about those already burning in other places around the world. These fires come on the heels of a hurricane that devastated the Caribbean but spared the East Coast of the United States—this time. Widespread and damaging floodwaters that plagued the midwestern United States for months have finally begun to recede. There's another article about the coming climate change apocalypse that has many people on social media participating in very agitated debates yet again.

Deaths from a mass shooting event make it the worst in the nation this year—so far. An infectious disease long suppressed in the general population seems to be making a comeback. There is a multistate recall on produce due to food safety concerns after dozens of people were hospitalized.

It isn't necessary to put names or dates or places on these events. They have become commonplace enough that at any given point in time, we can match these accounts to specific events that have occurred recently. It is against this backdrop that science communicators in the twenty-first century are doing their jobs, themselves working on emotional, contentious topics steeped in uncertainty and struggles for power. On top of that, those science communicators may have limited job protections, and most will not be in tenure-track positions.

This reality runs counter to the often-repeated idea that scientists must be incentivized to descend from their ivory towers to communicate with "the public." Instead, many science communicators are themselves the public—they live in the communities where they work, face the same issues and threats as their neighbors, and need support and protection more than they need incentives so that they can do the work that is most useful to the people in those places.

Bearing Bad News, Again and Again

"I live in Charleston, and people's feet are now getting wet," says Sarah Watson. As a coastal climate and resilience specialist with the South Carolina Sea Grant Consortium and Carolinas Integrated Sciences and Assessments, she works directly with communities to adapt to climate change. Watson is repeatedly in the position of being the bearer of bad news about how quickly increasing water levels are changing landscapes and lives.

"I recently had to give several presentations on the local effects of sea level rise to groups like planning commissions and homeowners' associations. Each time, at least half the people there were completely freaked out. I myself now get nervous having to say out loud what these climate projections tell us about tidal flooding in the region," Watson says. "It's hard information for communities to hear for what is often the first time, and it's hard for me to give the same bad news over and over again. News that I increasingly worry isn't *bad enough* to cover where we're headed but that I simultaneously worry sounds too alarmist for what people are ready to hear."

Before her current job, Watson spent years as an environmental journalist in New Jersey, where she was faced with a hurricane evacuation for the first time. The storm did not end up hitting her area particularly hard, but it piqued Watson's interest in climate adaptation. Following Hurricane Sandy in 2012, the self-proclaimed science and weather nerd jumped to the disaster beat full-time.

Watson recalls scanning some of the forecasts the week before the storm hit, projections showing it had an odd path and appeared to be making a severe left turn, heading straight for the Jersey Shore. "As I started to look at those models, I got a deep sense of dread that didn't go away for years."

Her dread was well founded. Within days, she was given the evacuation order she knew was coming. She was living on the third floor of a house on a barrier island and relocated to a motel farther inland and closer to the newsroom. As she left, Watson gave her keys to her neighbors, who were planning to stay at the time, and told them to shelter in her apartment if the flooding got as bad as predicted.

"That was my first experience of people that I knew really suffering from an impending disaster, and it affected me deeply. But I'm a communicator, a writer, and that was the only tool I had to help."

Back in the newsroom, she watched the tide gauges rise and listened to the police scanner. As the storm went on, even the first responders, normally talkative and confident, sounded somber and increasingly quiet. Then residents began calling the newsroom when 911 was too overwhelmed to assist everyone. Watson says she had no idea what to tell people; she had never trained as a first responder. But she tried to help where she could.

The newsroom stayed open, and when the power went out, she worked by candlelight with no power until relocating to colleagues' homes that still had electricity. Earlier that year, there had been a derecho, an intense and drenching storm with seventy-mile-per-hour winds, which Watson says sent the region "into the Stone Age." The electrical grid went down, and like so many places facing repeated disasters, the area had recovered only months before Sandy hit.

In the end, Hurricane Sandy caused an estimated $50 billion in damages and destroyed thousands of homes, and although the numbers remain contested, close to one hundred people died in the United States alone.[1] Watson says that she had intense guilt for many years after the storm, wondering what she could have done differently to be more effective at getting people out of harm's way.

In the midst of Hurricane Sandy recovery, which Watson spent more than a year covering, she was also preparing to go back to school to learn more about climate change and what she was witnessing. She notes, "It wasn't until I was back in school and learning more about human behavior that I started to let some of my guilt go."

Her graduate work in planning and policy at Rutgers led to her current position in extension—a system of professionals who provide locally grounded technical expertise as part of the Sea Grant or the cooperative extension system in the United States—which was a natural fit for the kind of applied scientific work she wanted to do. Only partly joking, she says she now wishes she'd also pursued a social work degree.

With her extension position in South Carolina, Watson has been working to make Charleston home. When she describes the smell of the South Carolina coast, her face lights up. "There is just this incredible fragrance the salt marsh has that draws people to it. I'm an ocean swimmer, and this place offers me tranquility." However, she says, "The irony is that the place I've chosen is shifting drastically at the same time."

She finds herself with the same wet feet as the people she works with. "I think that can be something that people don't always understand. In extension work, we live in the communities we work with, and that changes things. I'm not making predictions from a faraway office—I experience the same things everybody else experiences. And I have to look people in the eyes every day; I am accountable to them."

Watson notes that scientists and decision-makers spent decades talking about climate change as an abstract, distant problem. Today, she worries that current estimates of sea level rise, and the related advice she is asked to deliver about how high to raise a given building or piece of infrastructure, have in fact been too conservative. Not to mention the difficulties of addressing the many inequities inherent in disaster preparedness and recovery. On many days, she feels inadequately prepared to communicate the risks that she and her neighbors are facing. She is not alone.

To appreciate why traditional science communication methods are running up against big limitations in the face of these contemporary challenges, it's helpful to take a step back and understand a bit more about the evolution of the field and its emergence as a profession in its own right.

The Rise of Science Communication as a Profession—and Its Lingering Debates

The professionalizing field of science communication is still young, but it is not new. The vast and global history of science communication[2]—as long as there has been science, people have been communicating it—is best left to qualified historians, but reviewing some recent highlights that influence the central arguments of this book is worthwhile.

One early reference to science communication as a profession in the United States was published seventy years ago. In 1951, William F. Hewitt Jr. published a compelling article in *Science* presenting the need for a new focus on science communication. He argued that to open the channels of communication between scientists and the public, the large quantity of scientific information being produced needed to be organized, a concern that has only increased.[3]

Hewitt proposed that this work be undertaken by a group of scientist-communicators who would be recognized at a professional level comparable to that of academic researchers. Although the function he described was closer to how we might think of librarians today, Hewitt's idea of a field of professional science communicators was certainly prescient.

Since that time, science communication as a field has grown and remains somewhat fractured. For some, science communication means informal science education done in museums and at aquaria. For others, it comprises outreach to K–12 students in person and online. For still others, it's about podcasts or media interviews or who will be "the next Carl Sagan." There is also plenty of brilliant science communication happening on social media by a growing number of influencers who develop Twitter threads and games, Instagram stories, and YouTube videos that are truly and creatively engaging.

In addition, the US Cooperative Extension System associated with land-grant universities has one hundred years of institutional experience in science engagement. For those not familiar with the model, cooperative extension is integrated into the land-grant university system and exists at universities in every state across the United States and in several territories.[4] The system was designed based on the recognition, beginning in the early 1900s, that scientific research should be developed with and communicated to the people whom its results affect. There is no doubt that the history of land theft and racism that underpins land-grant institutions[5] must be addressed along with their legacy[6] to ensure that those institutions benefit all residents and communities in the United States. However, partly because of these issues, cooperative extension does provide useful lessons for community-engaged research and communication.

Furthermore, community or participatory science—previously called citizen science—can be yet another arena for science communication practice. Science writing by professional journalists is a related pursuit, though usually considered a form of communication different from the type scientists and professional science communicators engage in.[7]

In fact, one of the biggest challenges, and promises, for science communication is the variety of expectations and backgrounds that people bring to it. Many contemporary efforts are attempting to bridge the diversity of fields—including everything from psychology to journalism to data science—that science communication both draws from and contributes to, and there is a lot of work yet to be done.

Science communication itself can be broadly defined as "communicating science with nonexperts," whose ranks we all fill most of the time. It can go by other names as well—outreach, engagement, public understanding of science—terms that, while not synonymous, can refer to a similar set of efforts. While the precise definition can be endlessly argued, some contend that there are best practices that cut across fields. For example, much has been researched and written about what is known as the deficit approach to communication.[8]

The deficit model assumes that information is what is lacking in getting to any given outcome, whether that be leading people to take action or behave in what is considered an appropriate way when it comes to diet or vaccines or any number of subjects. However, research has shown that a dearth of information is rarely the primary barrier to action. Yet many experts continue to use this model, and some vociferously advocate for it and its cousin "Just the facts, ma'am." This debate and others like it will undoubtedly continue well into the future.

Discussion about science communication from previous decades tended to focus on the "newness" of the field. At this point, however, it is clear that there is not only a long history to science communication that extends beyond contemporary practice and the scope of this book[9] but in fact an overwhelming amount of information on how to communicate science. The much bigger job for science communicators, therefore, becomes discerning what applies to any particular context and working to fill specific gaps in their practice.

Advice on communicating climate change, for instance, is readily available. Some of it comes from climate scientists,[10] some from researchers,[11] some from advocates,[12] some from institutions.[13] It's coming from people and organizations in different fields ranging from climate communication[14] to psychology[15] to public health[16] and more. There is important work being done to illuminate whom climate communication and activism are done by, for, and with.[17] Some is based on research, some is based on practice, some is theoretical and based on what has "worked" in other arenas. Even wading through different perspectives on climate change itself necessitates more than passing interest, and while some people welcome the plurality of knowledge, others argue for better boundaries around what knowledge is truly valuable for climate communication.

Indeed, when getting science communication advice, it is helpful to ask where it originated, who is dispensing it—and to what end. How any given practitioner prioritizes their approach is determined by training, communication style, and audience. In other words, it's mostly about context.

The Challenges of Science Communication as Performance

Although it is difficult to strictly define science communication, some major assumptions have tended to guide the field over the last few decades. These generalizations overlook some specifics, but they remain useful nonetheless. First, there has been an often implicit belief that good science communication hinges on a performance by the communicator—the "sage on the stage" or in front of the mic or TV camera—who has largely been assumed to be a tenured or tenure-track faculty member.

This premise leads to the idea that good science communication is about the expert mastering such individual performance techniques as presenting information with meaningful graphics and developing a succinct message[18] to get the attention of elite journalists and policy-makers. It then follows that individual incentive and reward[19] for doing this "extra" work are necessary because they are external to the true research and teaching careers of academics. Both explicitly and implicitly, the focus on performance has meant armoring scientists with repeatable messages and emphasizing scientific authority, which have had the effect of creating distance rather than enhancing connection.

"The paradigm of science communication has largely been about the appropriate presentation of scientific authority," says Sarah Myhre, a climate scientist and executive director of the Rowan Institute, "which is about divesting from your own mortal and emotional and human connections. You are forced to perform respectability, to posture, and when you try to critique that posture or even just do things differently, you become the problem instead of the thing that's actually a problem. It's a losing game."

There are obviously many ways to think about performance,[20] including some excellent examples described later in this book that expand the idea of what science communication performance can be. However, in the normative science communication realm, it would be hard to deny that the primary thrust of science communication training over the past couple of decades has been helping individual scientists to become experts sought after by decision-makers and the media. Which has, of course, been easier for some figures, depending on many variables including identity factors such as race, gender, and sexuality; age; geographic and institutional location; and specialty.

This focus on science communicators who fit a certain profile—generally white, male, heterosexual, and cisgendered—has also served to decenter the knowledge of communicators who do not match that popular conception of scientific authority along with the needs of a wide swath of people and communities who do not see themselves represented by mainstream science communication efforts.

Although this performance focus is beginning to change, it is still the dominant paradigm first encountered by many students. After all, the primary people talking about science communication for the last few decades have been traditional academics, who are used to teaching in classrooms, and science writers, who are used to explaining complex subjects to nonspecialists. Many groups, including professional science communication training organizations and communications offices of scientific professional societies, have focused almost exclusively on connecting scientists with journalists and decision-makers.

Research communications expert Bob Lalasz refers to this phenomenon as the "big-media-science-communications-industrial complex" that binds scientists and journalists together in a stream of articles covering single studies. "The science-media-industrial-promotion complex has selection bias toward stories of discovery, scandal, and academic pissing matches," he says. "Those stories fuel an impoverished public understanding of what science is and can contribute—it's all about heroes, villains, and summarizing the next new paper instead of the larger questions of what we now know and what our options for action now are when new scientific knowledge emerges."

While that work connecting scientists and journalists will continue, it has led to a preconceived idea of what science communication looks like and the kinds of skills that are needed. Talking has been more valued than listening. Leading more valued than following.

Furthermore, the tight coupling of science communication and journalism has meant that the changing nature of journalism, at least during the last decade, has had a particular impact on science communication. There has been a major shift in the number and types of media outlets that cover science topics; many local outlets have shuttered, leading to a disruption in science communication practice as well. Simultaneously, science communicators gained access to new platforms such as blogs and social media that have allowed them to develop their own communication content and communities. Although being dependent on social media platforms has presented its own problems, such as online harassment, it has also clarified how interlinked science communication and journalism had become.

In addition to communicating with journalists and policy-makers, there has been a major focus on "communicating to the public" and "knowing your audience." While such framing is useful at times, interrogating those frames is also useful. First, because for most topics, everyone—including scientists—is the public. Understanding a specialized niche of science does not render one expert in all others. Taken that way, the dichotomy between scientists and "the public" holds less meaning.

"Audience" can be a similarly challenging concept. In the science communication context, it can bring to mind a passive group of fairly nonspecific people waiting to absorb whatever is performed for them. Communicating well then becomes a trick to understand or empathize with "them" well enough to have what is conveyed make sense or cause a behavior change. However, even defining a "them" requires a level of compartmentalization that is certainly simplistic and sometimes counterproductive. Again, there are more nuanced ways of interpreting the term *audience*,[21] but they are not the ones that make it into the majority of science communication training efforts.

Therefore, it is worth asking what science communication decoupled from a narrowly defined "performance for an audience" might look like. And there is no better place to turn than to the people doing that work in ways that might seem surprising.

Money Flows In and Water Flows Out

Like Sarah Watson, Diamond Holloman was deeply affected by Hurricane Sandy. Holloman, now a doctoral candidate at the University of North Carolina at Chapel Hill, says she can describe her research in these five words: "Hurricanes suck. Let's do better."

She also said of her experience with Sandy, "I was attending New York University at the time, and I lived in a dorm that was only blocks away from one of the evacuation zones. In tall buildings, when the power goes out, so does the ability for water to reach the upper floors. Of twenty-two floors, I was a resident assistant on the eleventh. We worked overtime, keeping the situation and residents as calm as possible, twenty-four hours a day, three days in a row."[22] She also began to notice that the disaster affected those less affluent differently, and she started to see the systemic barriers involved in disaster recovery.

"It got me interested in how we think about recovery and who gets access to resources, where those resources are allocated, and the moments that we don't talk about in mainstream coverage. Recovery takes longer than a couple of days or a few weeks. It takes years—sometimes decades. What I saw, heard, and learned during Hurricane Sandy were only things I had read about following Katrina. To see them play out in real life—that sparked something in me. And I've been following that spark ever since."[23]

While Holloman started graduate school thinking she was going to research Hurricane Sandy recovery efforts, after she moved to North Carolina, another disaster struck. This time, it was 2016's Hurricane Matthew, a storm that stalled over land and led to widespread inland flooding and wind damage. Two years later, Hurricane Florence hit the same area, long before the community had a chance to recover from Matthew.

One of the areas hardest hit by these hurricanes was the city of Lumberton, located in Robeson County, in the southeastern part of the state. In the county, Holloman observes, Black and Lumbee populations are the majority. Holloman's preliminary research indicates that federal aid and other recovery resources that go into Lumberton tend to exacerbate the problems of structural racism, noting that "Money flows in so that water flows out" is a common saying that conveys the idea that wealth ensures the impact of flooding is minimized. As an example, she says, "Affluent families, which are often white, own property that is more valuable and therefore tend to receive more money for recovery." That kind of inequality is compounded when people with hourly jobs, who may not own property to begin with, have to take leave from work during and after disasters.

"At the same time, these communities are resilient," says Holloman. She says that the Black community in the area has done effective grassroots organizing, building on efforts around other topics. Despite being able to advocate for what the community needs, it is still difficult to actually get recovery resources.

As she has worked throughout Robeson County, Holloman has found a variety of ways to not only communicate her research process and results but also act on what she is finding. For example, with university collaborators, Holloman is focusing on mold, which poses a major postdisaster health risk. Going beyond traditional science communication techniques such as flyers, pamphlets, and similar outreach, they are creating a tool-lending library to support residents in reducing mold and excessive moisture in their homes.[24]

Along the way, she has also engaged with a variety of personal and professional questions. While she is enrolled in an ecology program, her advisor is a geographer, and Holloman says she felt well equipped to understand her research as highly contextual and place based as opposed to broadly generalizable. But even with that outlook, she's not always sure how best to navigate questions such as whether it's okay to volunteer for or donate money to grassroots organizations that she's researching.

"You want to help the group, especially if you're asking them to give you their time or recount traumatic events to you. Most of the time, organizations are happy with me giving my time, but sometimes you want to help someone get their refrigerator working or find a new bed. How do you ask people about their lived experiences after disasters and then ignore the struggles they tell you about?" she asks. "What are the ethical dilemmas of profiting professionally from their stories?"

As prescient as Hewitt was about the place for a group of professional science communicators, he might be surprised by the very real predicaments faced by that group today. For some readers, it might be challenging to see what Holloman is doing through a science communication lens, and yet this grounded, creative, responsive way of working in partnership with communities merges both historical and future pathways in the field[25] while also raising new quandaries.

Science Communication and Advocacy: A Puzzle for the Ages
As Watson's and Holloman's experiences indicate, the issues facing science communicators today are forcing a reckoning with communication practices. When talking with people who have experienced ongoing disasters, odds are that messaging or framing efforts, curated slide decks, and brilliant thought leadership will just fall flat. Instead, there are deep questions to be explored regarding humanity, subjectivity, ethics, care, and so much more.

It is no surprise then that long-standing debates about when science communication crosses a line into advocacy—and whether advocacy is appropriate or even appropriately termed in this context—loom larger than ever and easily become entangled with matters of academic freedom and free speech. While the stakes feel high, the circumstances science communicators are confronting today are not entirely new.[26] To be sure, there have been critical moments throughout history where major societal transformations have forced changes in the sciences and science communication.

A case in point is the Nature Conservancy, which was founded in 1950 after a decades-long discussion within the Ecological Society of America led some members to ultimately decide that the society was not being proactive enough about what its research on land degradation was showing.[27] Similarly, the Union of Concerned Scientists formed in 1969, when faculty and students at the Massachusetts Institute of Technology set out to voice concern about how their research was being used by the US military.[28] Public Employees for Environmental Responsibility was established in 1992 to support scientists and others working in government agencies who felt their employers were not always acting in the public interest when it came to protecting public lands, enforcing environmental legislation, and the like.[29]

A web search for *science advocacy* will return any number of results for articles attempting to parse research from communication and communication from advocacy over several decades and in various scientific subfields. Young academic scientists are still exposed to the idea that advocacy is simply not something they should do if they want to remain in the good graces of the scientific community—at least not until after they receive tenure—and this admonition tends to spill over into science communication as well.

It may be that for some fields and people, the conventional wisdom makes sense, but for many others, it simply does not feel like an option. Climate scientist and communicator Kate Marvel of Columbia University captured an aspect of the problem when she tweeted, "On one hand, being a vocal climate science communicator can be detrimental to your academic job prospects. On the other hand, the past five years were the hottest on record, Mozambique and Nebraska are underwater, and California fire season is now year-round."[30]

Indeed, what the debate over science advocacy has regularly glossed over is that there are at least as many ethical concerns with standing on the sidelines as there are with engaging, that those concerns loom larger for some individuals, and that taking stances can sometimes defy what different constituencies, ranging from university administrators to government agency staff to communities, expect of scientists and communicators. Myhre, the climate scientist, posed a serious question in a *Guardian* article: "Our job is not to objectively document the decline of Earth's biodiversity and humanity, so what does scientific leadership look like in this hot, dangerous world?"[31]

For some, that leadership looks like being transparent about how close they are to any number of topics. Sara Kahanamoku, a doctoral candidate in paleoecology at the University of California, Berkeley, wrote about the controversial Thirty Meter Telescope proposed to be built on Mauna Kea. "To me, this debate is not about science vs. culture: in my practice of science, the two are inextricably linked. I am kanaka ʻōiwi, and I do science because I am Hawaiian. I research out of aloha ʻāina, a deep familial love for the land. My cultural upbringing allows me to walk in the space between Western science and traditional ways of knowing, a duality that enriches the questions I ask and the techniques I use to answer them."[32]

These are real situations that scientists and communicators find themselves in when trying to do their work and in dealing with the constraints placed on them by various institutional norms, as well as by laws and rules related to advocacy and lobbying that come from a variety of places, including specific academic institutions. While a practitioner can, and should, speak with relevant government relations offices, it is also worth noting that these rules and their interpretations are themselves quite subjective, and science communicators can find themselves in gray areas.

Where Institutional Demand Meets Individual Risk

Hewitt's point about the need to see science communication as a profession hits home when you consider not only methods but also the actual people doing the work and how they are positioned. In previous decades, there was an almost singular focus on science communication as the domain of research faculty, something they did "on the side," protected by tenure. Relatedly, there continues to be a tremendous focus on how to incentivize science communication for faculty.

This is valuable work, and the debates are certain to continue apace. However, the more the field grows, the harder it becomes to keep up with both research and practice in science communication. Not only that, but good science communication takes time and skill, mastery and artistry. It only makes sense that a professional class of science communicators is establishing itself, driven by individual desire, a changing job market and labor force, and institutional demand.

Indeed, science communication, and public schola[rship] broadly—in which scholars from all fields are encouraged to sh[are] work outside of academia—is becoming a considerable priority for [many] institutions, from universities to nongovernmental organizations. C[er]tainly, the public information office function in many organizations has traditionally been a driver for getting information out about the work that they are doing. From that perspective, communication was often presumed to be a universal good for institutions.[33]

However, what is missing from too many discussions today is the risk involved in communicating science, especially for the individuals doing the work. Overlooking hazards is once again a holdover from focusing on science communication as the domain of research faculty and specifically, culturally authoritative, tenured faculty members at universities with strong academic freedom protections and administrators willing to back them. As more science communicators work in contentious situations, the waters get muddier and muddier. This is particularly true for those working on controversial matters who are in already vulnerable positions owing to factors that include identity and employment status—which place them at more risk when it comes to critique from both inside and outside their institutions.

Indeed, as the following chapter will demonstrate, the people leading science communication efforts at a professional level are increasingly not in protected class jobs. They can easily be penalized, including by their own institutions, for speaking about high-stakes subjects in ways that might upset various constituencies. Therefore, extending protections to people doing science communication work without the possibility of tenure—at universities, in government, in the nonprofit sector, and beyond—should be at least as big a concern for the scientific enterprise, and society at large, as offering incentives and rewards to academics.

tion landscape in the United States has
two decades, and practitioners are deal-
ntentious issues that must be navigated

ation practices focused on delivering
from an elite scientist to an elite audience
not adequate and can be harmful in terrain where emotions,
conflict, and power must be traversed.
- Science communication practitioners are dealing with a wide variety of issues that have not been part of the traditional discussion in the field, and that reality should lead to an adjustment in advice given to science communicators.
- Science communication practitioners are increasingly in non-tenure-track positions and are not protected when working on contentious topics, necessitating a new emphasis on protecting practitioners versus incentivizing the unwilling.
- These changes invite new approaches to science communication practice, including directly addressing power and accountability.

Questions for Exploration
- What issues did reading this chapter raise for you?
- What does science communication mean to you?
- Where do emotions, conflict, and power struggles occur in your science communication practice?
- Are there places in your own work where you've seen a gap between the traditional advice given about science communication and how it works in practice?
- How might institutions begin to protect science communicators, particularly those without tenure?

CHAPTER 2

Science Communication Careers Today

SCIENCE CAREERS OVERALL HAVE SHIFTED DRAMATICALLY during the last decade, most pointedly in the aftermath of the global recession in the late 2000s. The fluctuations in science career options now appear to be a major driver of the changing value and practice of science communication. While there are certainly other elements that play a role—including the evolution of the internet and widespread use of social media—jobs are a key factor.

Only 20 percent of PhDs in ecology,[1] for example, will go on to academic research careers. More broadly, according to the National Science Board,[2] more than 50 percent of doctoral holders in science, engineering, and health are employed in professions other than academia within ten years of graduation—a trend for at least the last twenty years. Many of the rest will not go on to research careers and will instead land in practice-based careers, where desired skills—many of them communication related—differ from those normally taught in research-focused graduate programs.

While conversations about early shifts to careers outside academia are common, there is little to no conversation about support for those scientific careers through the mid- and late-career stages. As the narratives in this chapter indicate, a full-spectrum strategy that addresses not only graduate training but also advanced careers is called for.

An Uncommonly Common Career Trajectory

In the middle of her doctoral program in environmental chemistry at the University of Maryland, Marvourneen Dolor had a not-uncommon change of heart. She wanted to leave with her master's degree and pursue a public policy doctoral degree at another university. But, she says, "My advisor convinced me to stay and finish my technical degree and suggested that I consider a postgraduate fellowship as a way to pursue my policy interests instead."

Dolor ultimately agreed with the guidance and after graduating went on to complete a Sea Grant Knauss Fellowship at the US Department of Transportation. "I'm glad I followed his advice," she now says. "My doctoral degree gave me the confidence, and skills, to perform some key roles for my fellowship."

She quickly realized that one of the biggest assets in a position outside academia was her competence as a science communicator. Dolor found the kind of writing she was good at to be more welcome in a nonacademic setting than it had been in her doctoral work. "My writing muscle is very succinct. I talk in miles and I write in inches. In school, that meant I had to work at providing tons of context and caveats, but now my ability to distill information into a paragraph or set of bullet points is a huge asset." In addition, her talent for reading scientific and technical research papers quickly, and translating them for nonspecialists with ease, was more valuable than she had known.

Moreover, she says that the fellowship opened her eyes to "the small role science plays in policy decisions," which led her to learn a whole new array of communication skills. Like many scientists who move into government and other settings, she was quickly disabused of the notion that a lack of information or understanding was the key barrier to science-based decision-making. Instead, she had an eye-opening lesson in politics and the human relationships that drive it, which continued at subsequent positions with the Pew Charitable Trusts and the Great Lakes Observing System.

Today, she navigates the job market in much the same way most people outside academia do. She is constantly meeting new people and learning new skills. In just a few short years, she has pivoted several times, most recently to a better-paying and in-demand career in public interest technology. Dolor, a Black woman and foreign national who became a US citizen in 2017, says, "I think maybe it's coming from a developing country, being a woman of color in this country—I'm very focused on my earning potential, along with my quality of life, I'm not going to lie."

However, blazing a trail does take a lot of time and energy. Dolor has invested in building and maintaining her connections in numerous ways. During graduate school, her close proximity to Washington, DC, meant she was able to attend many events at the headquarters of the American Association for the Advancement of Science (AAAS) and other professional organizations. She has gone on to serve on organizational boards and be a founding member of the Ocean Collectiv, a network of experts that can contribute to various projects, and these pursuits help her to "always keep a pulse on things people are doing and thinking about."

Dolor's experience undoubtedly resonates with the many scientists who have moved out of academia and into a wide range of sectors. This large and diverse group of scientists and science communicators is exerting change both within academia and outside it.

The Changing Scientific Labor Force

The evolution of science communication as a field and profession cannot be decoupled from changing labor trends in the sciences. There are some science communication–focused graduate degrees in the United States currently preparing people for practitioner careers and for science communication research careers—and their ranks are growing. However, it remains that many people in science communication careers are first trained as scientists and then transition to science communication practice. Therefore, it is worth considering how to better prepare these students for the work they will do.

Although it is not yet commonly used in the sciences, the term *practitioner* is becoming more useful year by year. In the context of science communication, a practitioner is someone who practices science communication full-time or partially, either instead of or in addition to being a researcher of science communication or any other field. In much the same way that a medical researcher does not do the same work as your family doctor, science communication practice is distinct from research.

However, even in tenure-track academic positions, a successful career today contains many elements that have not traditionally been a part of research training, including science communication and engagement. For some tenured and tenure-track academics, that means learning these new skills on their own. For others, like Gannet Hallar, an atmospheric scientist at the University of Utah, it has led to an interest in partnering with professional communicators. "It's difficult for researchers to be good at so many different things. If you're somebody who wants to be a researcher and communicator, I think that's great, but there should also be a way forward for people like me, whose real strength is in research, to collaborate with the people who are good at communication."

Discussions about practice-based careers—also known as alternative or diverse or nonacademic careers—have been happening for decades, but they have ramped up in recent years. Driven by student demand, many campuses, professional societies, and other scientific organizations regularly hold sessions, panels, and workshops on how to navigate practice-based careers. Both by choice and in response to a changing job market, more and more scientists who train in research settings are becoming practitioners. They are joining a large field of professional communicators, writers, artists, and other science communicators with varied training and background.

So, not necessarily by design, universities are producing doctoral-level scientists who are released into practice-based careers without practice-based training or experience in organizations that may have vastly different professional norms. Compare this with training for lawyers or medical doctors, which has clinical, or practicum, components during the education process. Students are able to apply what they are learning in school, and while training, they also learn what the actual work they will be doing is like and develop professional networks. That is not to say the training in either of those fields is perfect or an exact model for others; however, they are at least set up to train interested students for practice-based careers, something currently lacking in science communication.

Training Practitioners

Disciplinary learning is undoubtedly at the core of what scientific training provides. In a 2013 research paper, Jessica Blickley and colleagues combed nonacademic job postings and interviewed practitioners in the field of conservation biology to figure out what kinds of knowledge and skills employers were seeking.[3] At the top of the list were disciplinary knowledge and analytical skills, followed by things that are at least somewhat transferrable, like project management.

Dolor's experience—being valued for her ability to "translate" research for nonspecialists—is common. In the field of ecology, translational ecology has arisen as a distinct emphasis,[4] modeled on efforts in fields like public health. In addition, programming, field experiences, and grant seeking are all transferrable to practitioner careers. It is worth noting, however, that teaching these skills in ways designed to support students looking to pursue practice-based careers would make for a more seamless transition. For example, scientific writing for journals or reports does not always easily transfer to other kinds of writing, but it could be taught in a way that does.

Many skills needed in practitioner careers are rare or deemphasized, even actively discouraged, in research training. Not surprisingly, interpersonal, communication, conflict, and cultural skills—infrequent foci in graduate seminars in the sciences—all showed up in the Blickley study as desirable. Beginning to teach to this set of skills is relevant not only to potential practitioners but increasingly to researchers and faculty members as well. Indeed, expanding traditional scientific training to include a broader set of skills offers many potential benefits for all trainees.

Practitioner training is common in many fields. Lawyers, doctors, and psychotherapists are all obvious examples of practitioners who are trained directly for their careers with requirements for clinical work incorporated into schooling, not to mention the continuing education typically required throughout those careers as a part of practice. Given the job market–related need and student desire for practitioner-oriented training, the fact that the sciences writ large have not caught up to fields with clinical or practicum-based models of education in any widespread and equitable way is worthy of discussion.

There are certainly complications to implementing the kinds of practicum-based models of graduate school that law and medical schools offer, the most obvious being the options for adequately compensated work. Teaching hospitals are set up to offer integrated clinical opportunities for trainees. Law students work on cases in school and benefit from on-the-job training during campus breaks. It is less apparent how sustainable models for this kind of practicum experience could become widely possible in fields with less money and opportunity to begin with. While some students secure internships and other on-the-job training in fields like science communication, the arrangements are frequently ad hoc and dependent on a supportive advisor and administration to make them work. In addition, the funding situation is case by case.

Another challenge for instituting practice-based training in science programs is that the people best suited to recognize the kinds of skills that are needed are generally practitioners themselves. While guest lectures by practitioners are common on campus, they are piecemeal and not frequent enough for students to make lasting connections, nor is a volunteer model entirely sustainable. Assigning these teaching roles to visitors also implies that their careers are lacking.

Incorporating practitioners into graduate programs has its own complexities. Leaving practice-based work for a teaching career can mean that valuable practitioner knowledge becomes quickly outdated. Finding employers amenable to half-time employees who do practice work and also teach can be complicated, but it is possible. Again, legal, medical, and business schools can provide some models for how to address these kinds of issues.

Taking a further step back, it is valuable to remember that science communication itself can be complicated for students to pursue in general. "Graduate students are actively discouraged from doing things that are 'unrelated' to dissertation research, in part because of funding but also because of time. The products of a science communication practicum aren't legible in academia," says Lucy Andrews, a graduate student in environmental science at the University of California, Berkeley. "You can't currently get a dissertation done with science communication products, only peer-reviewed publications. So if you go the science communication route, you're risking extending your time to finish your doctoral degree, which is already a long process and for which funding is tenuous and difficult to extend."

These challenges are worth addressing for multiple reasons. One is that there are many more scientifically trained people distributed throughout the workforce than are accounted for. While the sciences generally speak of people working outside academia as "lost" from the sciences, they are not. In fact, they are an asset. Creating a more cohesive and fluid relationship among scientists in different sectors is to the benefit of all.

The biggest impetus for change, however, is the fact that most current graduate training is not reflective of the jobs that many trainees will go on to do. Given the number of students who will be working outside academia or in non-tenure-track positions within academia, training needs to be rethought. However, altering training does have far-reaching implications for the scientific enterprise. For example, research labs depend on student and postdoctoral labor, and therefore truly addressing concerns about training outside of the university will mean attending to how scientific knowledge is produced.

Navigating Practice Work as a Graduate Student

Priya Shukla is a doctoral student at the University of California, Davis, where she is participating in the Sustainable Oceans National Science Foundation Research Traineeship program. The program was designed to expose graduate students to policy and management efforts related to their research, with the idea that what they learn will inform their ongoing doctoral research. Being a part of the program meant that Shukla recently spent a summer working for the nonprofit science communication organization COMPASS during her doctoral training.

"Early on, one of the faculty members that applied to bring the sustainable oceans program to campus decided that the best way to expose students to science policy and communication work was by embedding them with groups that specialize in that work. I've been able to benefit from that focus, as did the organization I worked with. The university funding was such that it didn't cost COMPASS funds to support me. I wouldn't have had this opportunity otherwise," says Shukla.

It turned out to be a win-win for both groups:

COMPASS had been working to hold an aquaculture roundtable, and aquaculture is what I research. My needs and theirs lined up. I was able to put on my empirical hat and offer my aquaculture expertise while also learning a lot from them about how to organize a successful and productive conversation.

I learned some things that surprised me. I would not have realized how much effort goes into creating the space for people to have a truly productive conversation. For example, a big question for the aquaculture workshop was just who should be in the room, which is not something I've had a chance to think about a lot before.

Shukla continues, "Who do we not normally hear from? Who is going to work well together? How do we facilitate this conversation? You have to think about the flow of the meeting, about power dynamics, about the order of the talks. Then there's distilling the results of the meeting and trying to make sure that it addresses a need and isn't redundant with what is already out there."

Shukla says another big benefit of working with COMPASS was having the chance to do congressional visits with people who are well connected in DC. "The organization has decades' worth of relationships, and we were able to have different kinds of conversations than I would normally have access to. Getting to be more than a fly on the wall was special. But I also realized that I have a lot to learn. I know how to give an academic talk pretty well at this point, but I don't quite have the language for those policy conversations yet."

Now that she is back in her research program, she says the experience changed her views on science policy and communication significantly. "I'm back from having dipped my toes in the policy pool, and then I hear the kind of advice that academics who haven't done any policy work themselves give to students and am able to reflect on it more deeply."

Shukla says that she has always loved research, but she now has a greater appreciation for policy work and how it links to her research. "Like most of us, I've written papers where I say 'this research has policy implications' without being specific about what that meant. It felt pretty hollow. I now feel much better equipped to start to think about those specific implications of my research. I couldn't have learned that by sitting in front of my computer looking at models."

Already a successful science writer with a regular column in *Forbes*, Shukla is excited to keep learning about new career possibilities. "One thing I'd never considered was working in philanthropy, but just being in these different settings and meeting people with science training doing so many different kinds of work is eye-opening."

Transitioning to Practice-Based Careers

One of the most concretely valuable outcomes of introducing a practice-based component to scientific training for students interested in science communication is that it allows students to connect with the people most likely to be their future employers and colleagues, as Shukla's experience indicates. It is undoubtedly true that a majority of job prospects, particularly in today's job market, materialize because of whom, and not just what, you know. Therefore, students who do not have practitioners in their networks will have a harder time finding satisfying careers and making it through the hurdles that the transition brings.

It makes sense then that the most frequent advice given to students who would like to transition to practitioner careers is to complete a science policy fellowship. There are many options now available, ranging from the AAAS Science and Technology Policy Fellowship to the Presidential Management Fellowship to the Sea Grant Knauss Fellowship that Dolor described. Increasingly, states are also developing fellowships—such as the California Council on Science and Technology Fellowship—that allow students to work in state legislatures. The AAAS Mass Media Science and Engineering Fellowship also presents an opportunity for students to spend ten weeks working with media organizations around the country.

These fellowships provide fantastic opportunities that have led thousands of scientifically trained professionals to move with relative ease into practice-based careers in government and beyond. However, focusing on these fellowships as the key, and sometimes only, piece of advice for students wishing to move out of academia can be inadequate. Many students, like Shukla, are increasingly interested in building skills *during* graduate school, something that practicum opportunities could help to provide. In addition, so far, the fellowship opportunities have been limited to government agencies, to the exclusion of nonprofits and industry.

More broadly, there is a discussion to be had about other routes to becoming a practitioner. Karin Tuxen-Bettman, a program manager at Google Earth, moved directly from a doctoral degree and a year spent on staff at the University of California, Berkeley, into her job with Google, where she has now been for a dozen-plus years. Having specialized in remote sensing and geographic information systems for her research, she could have taken multiple paths, including consulting or remaining with the university. However, when she saw that Google Earth was beginning to hire new staff, she knew immediately that was where she wanted to be and put many months into connecting with current employees until she was hired, she says.

Exposure to multiple tracks into practitioner careers can demystify and normalize the process for students. Connecting with the large group of already existing practitioners is therefore a key strategy for students and something that could be more systematically built in to graduate training, perhaps supported by a dedicated professional society or other network.

Navigating Early-Career Practice Work
Connecting directly with practitioners has worked well for Julian Reyes, who is currently serving as an AAAS Science and Technology Policy

Fellow in the Office of Global Change at the US Department of State. During his doctoral program in civil and environmental engineering at Washington State University, Reyes was fortunate to be part of a National Science Foundation Integrative Graduate Education and Research Traineeship program. That meant that during his training, he was exposed to a wide variety of career options that he says opened his eyes "in terms of what was possible with a doctoral degree."

One of Reyes's most rewarding experiences was spending a summer during graduate school as a science policy fellow with the US Global Change Research Program in Washington, DC. That, he says, is where he was exposed to science communication in a more direct way, learning how to communicate about climate change in particular. "It flipped things for me. I learned how valuable it was to connect with people, to listen, as opposed to coming in with a PowerPoint and an agenda."

That experience also made Reyes see that science communication opportunities often come down to luck. "I would have liked to have more science communication as an explicit part of my training. I think graduate students are navigating the field in a piecemeal fashion that is too dependent on circumstances and a sympathetic advisor. It's not particularly equitable."

Reyes also spent several years working with the US Department of Agriculture's Climate Hubs, first as a graduate student and then as a postdoctoral scholar:

> Starting AAAS and other fellowships, particularly straight out of graduate school, can be challenging for people. I am glad I've already been exposed to the different cultures of academia and the government, and to the politics, of course. I've witnessed firsthand how quickly a job can shift depending on who is in the White House. And it's also helpful to understand the basics of navigating the federal employment system and how people

actually end up getting into full-time positions. I was cognizant of what I was getting into, and I don't think that's always the case. It can be a pretty big learning curve.

When Reyes was interviewing in various offices as part of the AAAS fellowship application process, he knew to pay close attention to things like how many former fellows had gone on to full-time positions in the office. "It's not a guarantee of employment, of course, but it does indicate a willingness to at least consider fellows for permanent positions, which is important to me." Understanding how the system works gives Reyes a considerable advantage and is directly tied to his practitioner experiences during his graduate training. He laughs as he mentions that it also took him eight years to complete his doctoral work but says he wouldn't have done anything differently, because he is now well prepared to navigate this early-career stage.

Of course, he's already thinking about what his next job will look like. He's excited to be learning about international policy during his fellowship while simultaneously being open to what comes next. "It's a necessity in this job market," he says. "It's tricky to both try to put so much of yourself into a new position while also having to keep your options open for a more permanent position. Luckily, I have my trusted mentors and a large network of capable and supportive colleagues already doing this work to help guide me."

Supporting Practitioners

For too long, as previously stated, the conversation around science communication has been predominantly about efforts by research faculty, not primary careers. In fact, Tuxen-Bettman, who creates complex data visualizations and other interactive cartographic products for a living at Google, says she does not consider herself a science communicator. "I guess I always just think of it as something professors do," she says.

She is not alone. That perception is a big part of why there can be such enormous gulfs between science communication training, research, and practice. For example, training to give a media interview as part of an academic position is very different from the skills needed for a full-time science communication career.

However, to support science communication practitioners, one first has to be able to recognize them, which can be problematic when the field, and the nascent profession, is potentially vast. As mentioned in chapter 1, everything from informal science education to community science to data visualization, among many other efforts, contain science communication elements, even when that is not the main focus of the work. The organizations doing science communication work are diverse, and individual job titles themselves are all over the place.

In addition, there is currently no single professional society that science communicators belong to. There are many professional scientific societies with sections, chapters, work groups, and events related to science communication. However, navigating large, multiday scientific conferences where perhaps a handful out of hundreds or even thousands of presentations and posters are relevant does not yield a large return on investment in terms of either the networking or learning opportunities for many advanced science communicators. There are also many professional communication societies, but they are focused on communication more broadly. While some of that communication information is relevant, it is not so easily translated to science communication in particular, in part because of the unique concerns of practitioners, such as working on controversies that include climate change, vaccines, and pandemics.

Professional scientific societies could be doing more to better cater to practitioners, including by having practitioner tracks at conferences, dedicating permanent sections of journals or entire journals to practitioners, and encouraging mentoring programs. These kinds of efforts would benefit the societies themselves by increasing membership numbers, reinvigorating conference and journal content, and providing opportunities for collaboration and mentoring.

There are also more fundamental obstacles to supporting practitioners. When a practitioner is trained as a researcher, for instance, it can be difficult to completely remove the researcher from the practitioner. That is to say that research can still be core to the way a practitioner works. It can therefore be frustrating to come upon a question related to practice-based work—anything from the role of scientific information in a local conservation dispute to how to communicate with people who have survived a specific disaster—but not have a job description or resources that allow it to be pursued nor functional collaborative relationships with researchers who might be able to pursue the question jointly.

Those collaborative relationships can come with their own provocations, as the academic reward system does not yet lend itself to practitioner partnerships in a way that honors what practitioners bring to the table. It's difficult for practitioners to pursue research questions without access to the graduate students and postdocs so central to academic productivity. Even when it does all come together, finding outlets to publish in, particularly given growing journal fees, can also pose a problem.

In addition, establishing knowledge or expert authority can be difficult without a "professor" title or the "right" institutional affiliation. When talking to journalists, it can mean negotiating how one's role and title are presented, assuming journalists can find their way to practitioners at all. These barriers are exacerbated by issues of intellectual freedom and by aspects of practitioner identity, affiliation, and location.

Finally, advanced professional development programs for practitioners can be hard to find. Science communication training is currently designed almost exclusively for beginners, and those looking for advanced training must turn to opportunities that offer related communication or other skills but rarely specific science communication training content on subjects like misinformation, scientific controversies, and advocacy.

The truth remains that in many science fields, practice-based advancement paths are confusing and nondescriptive. In a tenure-track position, while still certainly complex, advancement is a relatively conventional path, with many others who have done it, or are also doing it, sitting in offices nearby. Correspondingly, it can be difficult to establish a salary baseline for practitioners. Finally, in current practice-based careers, the goalposts that make it clear one is advancing, or even keeping up, are simply not there.

Navigating Mid-career Practice Work
"I liked my work, but I was, like a lot of people my age, frustrated by my inability to move further in my position, or to impact the trajectory of our work, until the old guard retired. So, when it arose, I jumped at the chance to move into a position in another agency, one where I now have the opportunity to develop something new and effect positive change," says Mari-Vaughn Johnson. She is currently serving as the director of the Pacific Islands Climate Adaptation Science Center in Honolulu, Hawaii.

Johnson launched her career as a research scientist with the US Department of Agriculture's Agricultural Research Service but soon found that her voice and vision were not suited for the "traditional" research career pursued by her colleagues. Like many other "misfit scientists," Johnson found the AAAS Science and Technology Policy Fellowship to be a logical step to opening her career, serving at the US Department of State.

"The fellowship was pivotal in my career, partly because it showed me what I was capable of in terms of landing in this completely new setting and functioning relatively well," she says. "Working at State forced me to contextualize my scientifically based, physical worldview into a constructed, geopolitical global view."

The fellowship also drew on the budding diplomatic skills she had gained while researching grassland dynamics on ranchlands as a graduate student at Texas A&M. "I used to talk to ranchers in South Texas for hours, just standing around the backs of pickup trucks and never bringing up the words 'climate' or 'invasive species' even though I was addressing those things on their land. Instead, I spoke with them about their connection to the land, the value of restoring hunting opportunities. It helped me see there are lots of ways to connect with people."

Like Dolor, she quickly saw the complexities of integrating science into the policy- and decision-making process. "I'll never forget sitting in a room full of Foreign Service Officers when a scientist from another agency came in to give a talk. I was following intently, but then I looked around the room and everybody else was looking down at their phones, waiting for him to get to the point," says Johnson. And like Reyes, she realized that generating information was a small part of the work she would need to do.

After the AAAS fellowship, she returned to Texas to begin a position with the Natural Resources Conservation Service, a job she stayed in for nearly a decade. Along the way, she participated in the Global Young Academy, a program that allowed her to connect with other technical professionals all over the world. It was a pivotal professional development opportunity that she also recognizes as rare. "It helped me to develop a community, but I'm still searching for that elusive group of people that I can connect with professionally for the longer term. It's hard to find when you don't follow a traditional science career path," she says. "The community of people who sit between spheres is growing, though. Sometimes I think the people who are strongest in these roles wandered around and found their way into them rather than aimed to boundary span[5] immediately."

She found that mentoring a new scientist in her agency was helpful. "Mentoring her was great, but I think I learned as much from the experience as, or maybe more than, she did. Talking to her about how we shouldn't settle—should be brave and explore new opportunities instead of relaxing into known and stagnant patterns—drove me to jump ship and try something new, terrifying, and exhilarating."

Indeed, Vaughn-Johnson did what many people do to navigate mid-career transitions, which was relocate. Her current work in the Pacific Islands focuses on a coproduction model—in her case, connecting scientists with land use agencies and natural resource managers to develop science-informed climate change adaptation strategies. "This kind of 'coproduction' is a lot of relationship building and science communication, and in many ways we're still making it up as we go along. It can be uncomfortable, and it takes a lot of time because building trust is complicated."

As someone who has long connected with a wide variety of people around various topics, she feels well placed to do this work. "I've spent at least a decade cultivating these spaces 'in between' academia and practice, which can be both wonderful and lonely. There are so many tensions and opportunities between people like me and more traditional academics, and it does take a lot of work to navigate successfully."

Implications for "Academic Freedom"

It would be negligent to talk about practitioner careers without noting that this job shift—where more practitioner than academic positions are available and of interest—is changing not just the lives of individuals but the field of science communication itself. In some positions, particularly with government agencies, communication efforts are already tightly controlled. And, as mentioned earlier, for the many institutions that include nonprofits and universities that increasingly encourage science communication and public scholarship,[6] the emotional and contentious nature of many scientific matters renders the people doing that work vulnerable to everything from online harassment to job loss to loss of professional standing.[7]

In the past, it has been difficult to convey the multipronged effects of working on controversial material to people who haven't experienced them. In many ways, however, the coronavirus pandemic has laid some of these issues bare for more people. Think of the question, in the United States, of whether face coverings, or masks, could help to stem the spread of COVID-19.[8] What could have been a fairly straightforward scientific and technical matter turned into an enormous controversy that continues at the time of this writing.

Whether masks were useful seemed a relatively straightforward scientific question but was almost immediately mediated by the lack of availability of masks, which then played a role in the advice that was given about them. Many scientists and health-care workers who recommended them were chastised, with the controversies Anthony Fauci, the medical doctor who directs the National Institute of Allergy and Infectious Diseases, has faced serving as a prime example.[9] In subsequent weeks and months, after people started to make face coverings for themselves and others, it morphed into a politically polarized issue. Even today, when it seems evident that properly worn masks donned by a majority of the population can indeed help prevent further coronavirus infections, many refuse to wear them.

That is just one example, but many of the same dynamics play out in areas ranging from climate change to controlling large carnivore predation on livestock. These kinds of controversies and power struggles have far-reaching implications for science in general. After all, tenure was meant to offer at least some protection for people doing controversial work. Today, however, even within the category of academic jobs, fewer and fewer of them offer a tenure track, and increasingly many are not only grant supported, and therefore dependent on available funding, but also temporary appointments with a set term.

So what happens when more and more of the people leading the front-facing efforts of sometimes controversial communication work will never have the protections of tenure or, because academic institutions are not involved, the independence of academic freedom? Simply put, there is no way for a mass number of scientists to move into unprotected job classes to do controversial science communication work without causing major reverberations for science, a trend that few organizations are dealing with at all. If it is no longer the case that most of the people communicating controversial topics will have tenure, then the risks of communicating controversial information need to be transparently stated, and thought must be given to how to support and protect those doing it in new and creative ways. Today, this is a discussion that should be on equal footing, at least, with how faculty are rewarded for science communication work. While reward will undoubtedly continue to receive a lot of attention, it is crucial that it not continue to eclipse a conversation about the actual risks borne by the many whose jobs are precarious and what that means for communities and society overall.

"The assumption that science communicators will have tenure has led to an overemphasis on incentivizing science communication," says Sarah Myhre, the climate scientist from chapter 1. "What's missing is the back-end conversation that lets you know that, actually, 80 percent of the job opportunities that you might be interested in, those organizations are not going to be interested in you, because of your work as a communicator or public scholar. I have the same kind of job insecurity my blue-collar partner, who is a chef with a small restaurant, has. I don't have an elite job. I'm not in some ivory tower looking down on society and making pronouncements. I'm cobbling together a career."

Myhre goes on to say:

> It is a huge disservice to scientists to set them up to "communicate science" with YouTube channels and Instagram accounts and assume they will find a way through the job market based on being entertaining and personable. I have a lot of feelings about this because there was a time when I was that innocent. I invested time in this career—I'm not martyring myself entirely.

I chose this, and I need to make the best of it. But I believed in these institutions that were feeding me this line saying that if you get out there and communicate your empirical work, you will succeed—and it's just not true. Many of the things that will happen are predictable, and they're not success. You will lose opportunities. You will be cast as a political agent with an agenda. People will call your bosses and try to get you fired, people will hit on you and sexually harass you. Your professional paths will be foreclosed upon. And you will have to innovate your way out of the hole that you've dug for yourself.

Indeed, there is little discussion about how institutions can step forward and shoulder some of the risk involved in communicating controversial information or speaking up on controversial topics. While a science communicator risks anything from a basic reprimand to job, career, and even bodily safety, their institution can claim a large share of potential rewards in the form of attention-getting headlines and increased grant funds.[10] Right now, many individuals who do science communication, particularly on controversial topics, are effectively bearing all risk unless they work for a progressive organization attuned to the contemporary communications landscape—and provided they have leaders willing to back them. As some people are gravitating to more supportive organizations, others are finding individual solutions within complex institutional dynamics to create multiple layers of protection for their work.

As reported by the *New York Times*,[11] Patrick Gonzalez, the principal climate change scientist with the National Park Service, received a cease and desist letter from his supervisor after he testified to Congress about climate change in the national parks.[12] However, Gonzalez also has an adjunct faculty position at the University of California, Berkeley. He gave his testimony as well as served as a lead author on an Intergovernmental

Panel on Climate Change (IPCC) report with that affiliation, and the university supported him.

The agency's letter to Gonzalez claimed that his Berkeley affiliation was not separate from his employment with the federal government. Despite that warning, Gonzalez "disregarded the letter and spoke to Congress again, spoke to the media, and continued to work on the IPCC, in defiance of the National Park Service." He adds that "the National Park Service subsequently relented and classified my university position as an outside work activity separate from my federal government position."

These are gray areas with much subjectivity involved. Even with the kinds of institutional arrangements Gonzalez has, things can change quickly with new leadership and shifting priorities. Sarah Myhre looks toward other fields, including sociology and philosophy, to better contextualize the science communication landscape, saying, "The science community is like a little snow globe inside of the larger world. If you don't know you're in the snow globe, you think you're in a blizzard. We end up dismissing conflict as partisan or polarized, and we are missing the forest for the trees."

Sociologist Tressie McMillan Cottom began navigating the complexities of public communication as a graduate student and has continued to address them as a professor. She wrote of meeting a younger researcher: "Like many of us, she had heeded the call from universities for scholars to demonstrate their value to society through engaged 'public scholarship.' It's not an altogether altruistic call. Money is tight, tensions are running high, and higher education is making a case for its continued legitimacy to an increasingly disenchanted public."[13]

This is a common story within academic institutions today. She goes on to explain how meeting the call for increased public scholarship can affect individuals and their careers: "An ambitious, curious sort, my colleague acted in good faith with disastrous consequences. A rather innocuous Op-Ed had unleashed a negative response from colleagues and friends that caught her off guard," McMillan Cottom wrote. "This young woman learned that doing academia in public is risky. It's something I could have told her. I could have also told her that the risk is not the same for everyone."

It is worth reading McMillan Cottom's full essay and further work[14] for their rich and nuanced discussion, which encapsulates many of the pros and cons of visible public scholarship work. "The irony of good public scholarship is that when it is done well it will inspire strong reactions," she writes.[15] That truism reverberates across the landscape of public scholarship and science communication in the twenty-first century.

Zuleyka Zevallos, an applied research sociologist, developed a framework[16] that outlines steps individuals and institutions can undertake to address some of the risks involved in public scholarship, particularly for those who work on controversial topics and are marginalized by factors such as race, gender, sexuality, and age. She first notes that, of course, scholars have a responsibility to use their knowledge for the public good.

Simultaneously, Zevallos writes that institutions have a duty of care for their employees. "Institutions benefit from media interviews and social media work by academics and researchers. That means equity and diversity considerations must be undertaken to support staff when they are besieged by public abuse as a result of their justice education, publications, media appearances, and social media discussions. Media policies should not simply tell academics what to say, or what not to say. These policies must also clearly show measures and processes to protect the safety of staff and students when their lives are endangered."[17]

Zevallos goes on to provide a checklist that includes an extensive set of questions that can be used to review relevant institutional policies, such as prevention, harassment policies, and physical and psychological safety procedures. She stresses that understanding the range of threats and protections available in advance of problems, as well as evolving the approaches over time, is important.

The universities and colleges and departments that are keen to promote science communication and public scholarship have a long way to go in explicitly agreeing to share the risk when it comes to controversial work. It's one thing to ask for support and incentives for science communication work, another to have to fight to retain your job and your voice and sometimes even your career. Despite these challenges, it is crucial to note that scientists are trusted members of society, and ensuring that trust is not abused is paramount. This is a tightrope that must be walked. Science communicators need protections so they can speak about controversial matters because it affects society when they cannot; however, they cannot operate with impunity.

These kinds of conversations will be needed to guide the next evolution of science communication as a field and profession. As the next chapter will further demonstrate, many of the areas that scientists and science communicators find themselves working in are deeply emotional and contentious, and there is little guidance on how to navigate that terrain professionally.

Takeaways
- The scientific job market has changed, particularly in the wake of the Great Recession (2007–2009) and within the context of a highly uncertain economic situation today.
- Shifts in the scientific job market are leading many more scientifically trained people to become science communication practitioners with little explicit training.
- The assumption that science communication will be carried out by tenured faculty does not hold, and any advice on communicating must recognize the precarity, and creativity, of new positions.
- There are different concerns for all levels of science communication training and careers. While the conventional focus has

been on training graduate students, early- and mid-career science communicators need different kinds of support.
- The shift in who makes up the field has big implications for the kind of work that can be done.

Questions for Exploration
- What issues did reading this chapter raise for you?
- How have you seen the changing job market in the sciences affect the interests of trainees?
- What kinds of supports might be helpful to science communicators at all levels, from students to early-, mid-, and late-career professionals?
- How can you work in solidarity with others to protect people working on emotional and contentious topics?
- How might institutions better support science communicators working on emotional and contentious issues?

CHAPTER 3

Navigating Facts and Feelings in Science Communication

Scientists are encouraged to be as objective as possible by adhering to the scientific method. The overlap between science and science communication has led to a focus on objective or "neutral" information provision for decades. Yet the concept of objectivity has been debated from multiple angles, including by feminist scholars,[1] science and technology studies,[2] philosophy scholars,[3] and journalists.[4] Journalist Ramona Martinez argues that objectivity is a construct that has simply made normative the subjectivity of people in power, or more simply, "Objectivity is the ideology of the status quo."[5] There are also many commonsense challenges to objectivity, including the fact that science is done by human beings with complex subjectivities, which can become even more obvious when working on complex, emotional, and contentious subjects.

Rationality and objectivity—often seen as opposites of emotionality—are idealized in modern science, and there is a deep-seated fear that moving away from that norm will ruin the profession. However, science and science communication are stressful undertakings filled with emotions that range from the thrill of receiving a grant to the deep insecurity or unjustness of being rejected for a promotion. In addition, at a basic level, not recognizing the emotional context of climate change,[6] coral reef degradation,[7] food insecurity, and any number of other subjects can be stifling for individuals and prevent scientists and communicators from connecting with affected communities on many critical issues.

It can even become dangerous and harmful when scientists and communicators try to intellectually distance themselves from the emotional implications of research and communication efforts by ignoring feeling-based reactions, whether their own or other people's. Dismissing emotions can cause scientists and science communicators to miss out on the valuable role they can play as a source of information that can be used in robust knowledge production or, on the flip side, to justify practices that can cause harm.

In addition, as many scientists and communicators from marginalized backgrounds point out, that kind of distance is simply not possible or desirable when your community is being affected.[8] There are times when a commitment to objectivity, for instance, can be used to uphold systemic racism by making it "too political" to address. Concurrently, factors including race, gender, disability, age, religious affiliation, and marital status can make emotional and contentious terrain even more difficult to navigate as a practitioner.

Feelings about Food

Food is a vital part of life, and people's relationship with, and access to, it is a deeply emotional matter. There is no doubt that the term *emotional eating* exists for a reason. Yet, says Wei-ting Chen, the associate director of community partnerships in the Office of Community Engagement of the Stanford University School of Medicine, much research on the food system overlooks people's feelings about it as they go through the day, feeding themselves and those in their care.

"As I was working on my dissertation at Johns Hopkins, I read a research note from an interview with a mother who participated in a study that focused on how low-income caregivers made ends meet in the context of welfare reform. In it, the mother described making a fluffernutter sandwich every day for her young child as a token of motherly love. She was also concerned about her child's weight. Something about that struck me," says Chen. Although she had to look up what a fluffernutter is—a combination of peanut butter and marshmallow fluff on white bread—she quickly recognized that this brief note, which had not been part of any previous analyses, contained valuable information.

"It's obviously not the healthiest food, and the mom knew it, but there was so much more to it. The mother felt the need to express her love for her child by providing that food, and seeing that, feeling it myself, fundamentally changed the way I looked at my own work." As an immigrant and first-generation college graduate, and now doctorate holder, she was drawn to understanding the role that food plays in families and communities lacking resources to meet basic needs. When she initiated her own research interviews with low-income mothers in Baltimore, Maryland, she kicked the conversations off with what she viewed as a rapport-building question about their childhood memories related to food.

> I was utterly unprepared for my first interview with these women. I asked them to describe their childhood experiences with family meals. One woman described living in many different situations and developing different eating habits, and when I asked her more about it, her descriptions of abuse were overwhelming. She was sexually assaulted by a family member and as a result was placed in a different home, then went through foster

care, then off to live with distant relatives. And in every case, food was central to how she was treated. Learning every new household's way of dealing with food, having to try new things, or having food withheld—these were just some of the ways that food intersected with her ongoing trauma.

Chen says she did further research and found these kinds of interrelated traumas are prevalent, but she says, "I didn't realize how common it was; I just didn't anticipate that. It's not something that's talked about enough. I was reading a lot of papers on interventions, and no one was looking at household dynamics that way. It's acknowledged, but I still don't think we tackle all the ways that emotions affect how people deal with food."

A decade later, having worked on different aspects of the food system in a variety of positions, including as a health communications consultant, Chen says that this is still the case. In her previous job as a cooperative extension advisor in the larger San Francisco Bay Area of California, Chen says that when teaching nutrition education classes, the women who were responsible for cooking for their families would sometimes protest new recipes based on how they perceived their husbands would feel about the meals. "Emotions around food guide so much related to healthy behaviors, and I don't think we're effectively able to integrate that into a lot of food system work yet."

Being able to recognize her own emotional response to the emotions others had about food, and then to incorporate that into her research and practice, has been life changing for Chen. "For my advisor, I think the most important aspect of my doctoral research findings was that family routines matter and that social inequality makes it difficult for families to maintain routines, and that leads to poor health behaviors. That is important from a policy perspective. But the moment I saw that fluffernutter sandwich was a routine, I myself was moved to try to better understand the emotions that were involved."

Emotions Matter

As Chen's experience demonstrates, the emotions of researchers and practitioners, as well as those of the people they work with, are not broadly discussed. However, there are some areas where exploring them is more common.

For example, there is an active conversation surrounding emotions in climate science. In fact, when it comes to climate change, historian Naomi Oreskes of Harvard University has argued that scientists should express more alarm.[9] She recalled a conference presentation where an audience member stood up and said, "You are telling us that we have a very serious problem, but you don't sound at all worried. You don't even sound upset!" Oreskes argued that expressing concern would help convey the seriousness of climate change because it's difficult to get excited about something when the experts themselves seem dispassionate.

Some people are also concerned that stifling emotion could take a toll on scientists who work on difficult topics. David Corn of *Mother Jones* is among the journalists who have written recent articles about scientists experiencing "climate depression."[10] Indeed, over the past decade, scientists working on climate are themselves increasingly open to discussing the emotional toll they are experiencing.

Researchers in geography developed the idea of "emotional geographies,"[11] which addresses, in part, how emotions can shape how people interact with their environment. Farhana Sultana of Syracuse University writes that "resource struggles and conflicts are not just material challenges but emotional ones, which are mediated through bodies, spaces and emotions."[12] The several anecdotes that follow render this idea in concrete terms.

Sadie Ryan is a professor of medical geography at the University of Florida. As both a teacher and researcher, she has found that the strong emotions surrounding the subjects she works on—infectious disease and climate change—are coming up in the classroom more and more. "As we start to look at climate change implications for diseases and show the scary potential risks to places that haven't traditionally been in danger from, say, Zika virus or dengue, it starts to drive it home in a much more stark way than the older kinds of sadness I would see about a beautiful melting glacier we were never going to visit, or a starving polar bear on an iceberg. Seeing climate change as intersecting with a 'health crisis coming soon to a city near you' gives it a terrifying urgency instead," says Ryan.

She, like many scientists, can still be surprised to find herself navigating not just the information she is responsible for teaching but her own and her students' strong feelings about that information as well. "The flip side of these high-urgency crisis-level stories about disease outbreaks is that they disappear in a news cycle, and how we can address them in the future is ambiguous. Students in particular are then faced with a horrible sense of helplessness. I try to support them by getting them to be curious about how we continue to get on with our lives despite the risks, but it can be difficult for all of us because there's not really a map for navigating these emotional situations professionally as a scientist."

Mika Tosca, an assistant professor at the School of the Art Institute of Chicago and an affiliate climate researcher at the NASA Jet Propulsion Laboratory, directly addressed the material of one of her courses with a social media message sent to her students on the last day of class. Tosca wrote, "Aware that the material that we covered in class can be demanding and anxiety-inducing, I recognize the importance of ending the semester on a note of optimism. When you leave this classroom, and go out into the world, my biggest hope is that you can embody hope amid ubiquitous proclamations of despair."[13] She presents the problems of today as an opportunity for reimagining the future in revolutionary ways for these students.

Even field and lab work can come with strong feelings. Marilyn Myers, now retired from a career in government service in New Mexico and Alaska, revealed that her research with aquatic invertebrates made her extremely emotional. She says that while she loved sampling invertebrates from desert springs to document species diversity, distribution, and ecology, killing her research subjects in the process was difficult and never merited discussion in her training or professional life.

Myers says, "I know they were 'only' invertebrates, but I loved them, and it was especially difficult when I had to pour alcohol on living animals. Even now it is a difficult memory. I had compassion for them, and I grieved for them, but it all had to be buried. It's not something people talk about, especially when I was in school and doing my field and lab work. I just know for me they were special, and killing them was something that bothered me."

It is not only researchers and students and practitioners who are grappling with strong feelings. The people with whom scientists interact have them as well. For instance, Roger Betz, a cooperative extension agent with Michigan State University, spends a lot of time in intimate conversations over kitchen tables, helping families figure out how best to transition their farms from one generation to another. Writing about his work for the *New Food Economy*, Beth Hoffman says agents like Betz "become close confidantes and facilitators of conversations about farm transition and land transfer. These are some of the hardest conversations families face because they involve delicate and emotionally fraught issues of money, land, relationships, and death."[14]

At a larger scale, people's feelings about a variety of matters can have wide-reaching implications for public policy. Sarah Wolfe is an associate professor at the University of Waterloo. She researches how emotions, including fear, awe, and disgust, influence how people and institutions make decisions about water resources. Wolfe wrote about how disgust related to perceptions of water quality can lead to the overconsumption of bottled water in places where the tap water is generally safe. It also makes her own work on water challenging. She writes, "Human emotions are complicated, so talking about them is never easy. But given the many and multiplying stresses on our drinking-water systems, it's time to stop ignoring how powerful and universal emotions such as disgust both help and hinder our water decisions."[15]

This sentiment is backed by work from Parisa Parsafar, a developmental psychologist who studies how emotions affect children's perceptions of drinking water. She has found that emotions strongly relate to what kind of water children are willing to drink. In an interview, Parsafar said, "Early experiences of disgust and fear are linked to water contamination beliefs and water consumption behaviors among children. Although these might seem like individual issues, fear and disgust can have far-reaching consequences. Water contamination concerns could also reduce children's water consumption, which poses serious public health risks like dehydration and increased sugary beverage consumption."[16]

Karin Tuxen-Bettman, the Google Earth program manager from chapter 2, has spent years developing air-quality-monitoring tools for communities and says that air quality is a deeply emotional subject for the people she works with, as well as for her. "I grew up in Stockton, California, during the time when people would burn agricultural fields all around the Central Valley in the fall. And that smell just felt like home to me. But now I know that it was dangerous—today, it's strictly regulated—not to mention that these days when I smell smoke, my anxiety spikes because I worry that there's another big wildfire nearby." Her work with communities to better visualize air-quality data is something that she says "makes it harder to sweep under the rug because it's not invisible anymore."

Many of the emotions science communicators contend with can be difficult. However, there is also a desire to create, convey, and partake in joy. The Plant Love Stories movement, for example, was developed after a week of intense media and policy training as part of the Smith Fellows Program, says Mallika Nocco, a cooperative extension specialist at the University of California, Davis. "At the end of that week, we were exhausted. My colleague Becky Barak said she wanted to do something fun and joyful that would make people love plants. We wanted to focus more on 'plant appreciation' than science and just make it fun."

Nocco has been a longtime gardener and once wrote an award-winning "tomato love story" that got her tickets to the Tomato Fest, an heirloom tomato festival that was held in Monterey, California. "It just clicked for me that we could do plant love stories. Bonnie McGill, who is a scientist and artist, designed the logo and website, and we started collecting stories. My tomato story started us off, and then we asked people we knew to contribute. Plant Love Stories was born on Valentine's Day in 2018, and we now do yearly 'Plant Heart Art' artist collaborations to celebrate."

Nocco says that Plant Love Stories is a bright spot in her work. "It makes me feel joyful and light. It makes me feel like I can be funny. I have all these parts of my personality that I feel like I have to suppress for science, but with this I can be goofy and make puns and be a total plant nerd and it's okay."

From disgust to joy and agriculture to pandemics—including well-known controversies such as those over vaccination, genetically modified organisms, and fracking—emotions can influence everything from traditional field and lab research to teaching to community engagement and public policy. They are as much a reality as any other factor that plays a role in science communication practice and should not be underestimated.

Minding the Gap

How science communicators can best work with their own and other people's emotions is not always straightforward. Mila Marshall, a doctoral candidate in ecology at the University of Illinois at Chicago, says that science communication on topics that might test people's fundamental worldview takes both skill and compassion. "Take something like the Dust Bowl, which people commonly understand as an environmental disaster that happened seemingly by accident," she says. "But when you look at it in a sociopolitical context that includes slavery and Indigenous death and dispossession, you can see it wasn't much of an accident."

However, because this might be new information for some people, Marshall says she proceeds with caution. "I've learned to let people know where we are headed, because it can be emotional. We're talking about racism and hate, and people can easily withdraw from the conversation or go into their heads. So I ask people to take a breath and decide how much they want to feel into these issues, and whatever they decide is okay."

Marshall says she has learned to navigate emotions by thinking of them as invitations that can be accepted or rejected. "If you go to a party and someone asks if you want a drink, you get to say yes or no. In that same way, you don't have to accept every emotion in every situation," she says. "If you take the drink every time, you get cirrhosis of the liver. Instead, discern the best way to manage yourself while you're experiencing any emotion. Our emotions are real, but they aren't all we are."

Gaining that perspective has been a reclaiming process. "When I started graduate school, I was good at communicating. But as I moved deeper into my studies, I lost the ability to talk to people outside of science," Marshall says. "At one point I was discussing something with an ex and said, 'Show me your references,' and he looked sideways at me, and I was immediately like 'That was weird.'"

She started to feel academia was isolating her. "I felt like nobody understood what I was saying. My first response was to create this narrative of 'y'all are just stupid,' which was obviously incorrect. I value being able to talk with people and to translate information, and I was missing out on so much. I don't want to lose a new friend or ally because I'm only about science."

Marshall's current research is focused on nature and the Black community in Chicago. "There is so little information that is relevant and available for decision-makers. We have all these great academic institutions here, and yet we still miss the opportunity to increase the quality of Black lives on purpose. There's information and data that gets superimposed on Black communities but wasn't created by us, for us," says Marshall. "If I want to be a part of the 'supply chain of solutions,' I need to create information that's relevant to the community I belong to."

She applied this mindset in her work with a local nature center that has put a great deal of effort into culturally appropriate nature education. "Eden Place Nature Center is in a predominantly African American area with mostly renters on Chicago's South Side. There are high rates of unemployment and underemployment. Nature centers tend to focus on 'natural' landscapes—trees, birds, bugs, fish, watersheds, things like that. But that's not how everybody engages with the land," Marshall says. "That kind of typical environmental communication can push people away and make them feel as though you're not talking to people who look like them. It can create this perception that nature is only for white people. You have to know where you are, who you're talking with," she explains.

"These days, people end up wanting to connect Black people with the land through community gardens. But that creates more work for people who do not need more work. On top of that, a vacant lot might hold hundreds of kinds of living organisms, then you chop everything down and add fertilizer and weed suppressant so you can put in a garden with raised beds to grow lettuce, tomatoes, and cucumbers," Marshall says. "Why limit the landscape, and people, to food production? Black people deserve wetlands, prairies, and woodlands too. So at Eden Place, we created gardens, but we also created natural areas, all walkable in twenty minutes."

She notes that how they discuss nature is an important part of the experience. "We give Black people a chance to reflect on our relationship with the woods. I never wanted to hug a tree—I grew up with images of Black people being lynched. We have to talk about the woods in a way that doesn't make people feel anxious. That means conversations about the history of Black people in the outdoors and acknowledging how Black people can feel about science because of things like the Tuskegee Experiments and the harm that was and still is caused to our communities because of systemic racism."

Marshall stresses that emotional intelligence and cultural understanding are key in these kinds of conversations. "These are deep issues. People are sometimes relying on misinformation. When you're trying to talk about difficult subjects, you have to understand that you are asking people to acknowledge there is a chance that something they believe to be true, told to them by people that they love and respect, is wrong," she says. "When we untether people's realities and connections to other humans, we have to be careful. We can provide accurate information to people, but there also has to be somebody standing in the gap prepared to support them through their emotions around how it feels to be corrected and the embarrassment of going hard for what you thought was truth."

Whose Emotions Matter?

While emotions abound, as Marshall's experience indicates, working with those emotions can be fraught for many reasons. Geographer Sarah Ahmed has done foundational work here, particularly with her book *The Cultural Politics of Emotion*,[17] which among many valuable offerings, interrogates the way that emotions can be used against marginalized communities. In addition, whose emotions are considered—and where and how and why—is not one-size-fits-all. Race, gender, sexuality, and many other factors mediate how emotions are interpreted and acted upon. Emotions can even be used to justify "heroic" efforts on the part of scientists that may ultimately negatively affect the communities they seek to serve.

For science communication practitioners, expressing emotion can be complicated for any number of reasons. For example, because emotionality can be construed as gendered, what might be seen as overly emotional for some may be seen as brave candor for others. In addition, because anger is filtered through race and gender, it is more "allowable" by some than others. These are, of course, broad generalizations, but they illustrate the kinds of considerations of whose emotions count and how unequally "the rules" apply.

However, for many people, it is both impossible and undesirable to become untethered from their emotional ties to a topic or place or culture. As scholars from marginalized communities have long said, it is truly a privilege to be able to be seen as an "objective" observer of phenomena that directly affect people's lives. Even more so when it is your sense of self, your family, and places that you live and love on the line.

Lorena Villanueva-Almanza was a doctoral candidate at the University of California, Riverside, when she participated in a collaborative project examining the effects of a large dam proposed on the San Pedro River in Mexico. The river currently flows unobstructed into the Pacific Ocean at Marismas Nacionales, an internationally recognized reserve, in the states of Nayarit and Sinaloa.

Born and raised in Mexico City and having done extensive fieldwork in Baja California, Villanueva-Almanza has deep ties to ecosystems across the country and says she was deeply saddened by her work comparing dammed rivers to those that are free-flowing and felt she needed to speak out. She notes that the San Pedro River and its estuary provide important small-scale fishing grounds. Sediments that come down the river to ensure mangrove and other habitat for shrimps, oysters, crabs, and many fish species can be trapped behind dams.

She was stunned to see the effects of another dam in the region firsthand. "After five hours of driving, we arrived at the mouth of the long-dammed Fuerte River. We found an abandoned fishing camp. The silence, trash, solitude, and even the sparse vegetation was upsetting. People had left everything behind," says Villanueva-Almanza.

The team found that the beach was smaller, and there were many fewer species of plants and other mangroves to keep it in place. "Soon the ocean would drag what remained, and the river would not be able to catch up. The sediments would never make it to the coast. Perhaps it was the last time I would walk that coastline. It is hard to come to grips with what I saw that day and the losses for the fishing community. It was such an emotional experience that is difficult to explain with words."

These heartrending situations can truly leave scientists and science communicators, particularly those with ties to specific places and communities, at a loss for words. Simultaneously, these experiences feel too critical to overlook.

Responding to a Crisis in California's Central Valley

California's recent five-year drought hit the agricultural community in the state hard.[18] Perhaps none so hard as the small farms of Hmong and other Southeast Asian farmers that lease small plots of land in areas with declining groundwater levels, shallow wells, and outdated irrigation systems. The community had a brush with a mental health crisis—there were reports of a spike in calls to a suicide hotline as farmers faced fears of losing their livelihoods—as wells went dry and pumping costs increased.

For many small farmers, farming is part of who they are, despite the hardships. And the list of difficulties can be long: not owning the land they farm, decreasing acres of land to lease because of urbanization and the potential for growing higher-value crops than the Hmong specialize in, language and cultural barriers, and competition for groundwater. Yet many of these small farmers persist, growing an incredible variety of tropical and subtropical crops in California's temperate climate.

Ruth Dahlquist-Willard is a small farms advisor with the University of California Cooperative Extension program in the heart of California's

Central Valley. Trained as an entomologist, she works with mostly immigrant and first-generation farmers, of which there are thousands in her county alone. Dahlquist-Willard was new to her position as the emotional toll of the drought hit the farmers she works with. She and Michael Yang, a Hmong agricultural assistant also with cooperative extension who offers a lifeline to small farmers throughout the region, were receiving calls from farmers whose wells were drying up. They quickly jumped into action to figure out how best to offer support to the community during a difficult time.

As a first step, the team organized a survey for Hmong farmers with the support of Jennifer Sowerwine, a cooperative extension specialist at the University of California, Berkeley. They found that many could save money on their energy bills by switching electricity rate plans. Xai Chang, a Hmong farmer working with cooperative extension, then took the lead, giving growers tools they needed to find the best utility rate for their irrigation pumping practices. The team also found that a state grant program to help farmers upgrade their irrigation equipment could be a resource for tenant farmers and helped small farmers fill out the extensive forms needed to apply for the program.

"We held multilingual workshops to inform the growers of the grant process and get them started on their applications, then offered one-on-one assistance for completing the application and getting all the required documents together," Dahlquist-Willard explains. She organized a group of collaborators to assist growers. While county extension offices did not usually offer this type of assistance, the drought was a crisis that required a concentrated effort.

Preliminary results indicated that growers who received assistance reduced the water and energy used by their irrigation equipment, which helped their bottom lines. Although the drought is over for now, there are still many other impediments for small farms. Groundwater in particular is a looming area of concern. "There are a lot of open questions related to water availability for small farms. Especially for Hmong farmers, water is crucial to grow crops like lemongrass, luffa, long beans, and sugarcane. It was wonderful to see crops grow lush and green because we had water," says Yang.

"Southeast Asian farmers are tenacious and creative—they have already overcome incredible odds in adapting to a new culture and environment," says Dahlquist-Willard. She and Yang are committed to offering pragmatic support to Hmong and other small farmers through the many challenges that will undoubtedly come their way.

Incorporating Emotion into the Practice of Science Communication

When scientists and science communicators recognize and allow for the diverse range of emotions that can come up in practice, a key question emerges: What, if anything, should be done about it? Some hope scientists and communicators will convey the seriousness of a given topic and get others to act by expressing their emotions. Psychologists and others have been more focused on providing support and resources, like therapy and mindfulness, for researchers and practitioners.

For science communication practitioners, however, a different set of questions and concerns arise that have more to do with how to continue to work effectively, ethically, and sustainably. Chen let her own emotional response to a brief research note about a fluffernutter sandwich drive her research questions and infuses her understanding of the role of emotions into her communication and community engagement work. Marshall has developed ways to work deftly with other people's emotions when their views are contested. Dahlquist-Willard and Yang responded to a crisis in their own backyard by leading with a survey to hear directly from the people they work with about what they needed and then did that work. Most difficult of all may be the times when there is nothing to be done except sit with emotions that come up.

As emotions are embraced rather than pushed to the side, science communication efforts can lead to, get caught up in, and even cause conflict. Conflict is an ongoing presence in many scientific practice fields. Indeed, one of the biggest challenges of being present to emotions in a group setting is that they can simmer into a cauldron of collective emotions: sometimes joy but also grief, trauma, and anger. Navigating the conflict that can come with this kind of engagement can be hard but necessary work.

Bearing the Conflict in Fisheries Management
"It was a formative time for me as a scientist, because it was the first time I faced the industry," says Sarika Cullis-Suzuki, a marine biologist and science communicator. Just after finishing her master's degree at the University of British Columbia, Cullis-Suzuki was invited to present her research results at the United Nations (UN) headquarters in New York as part of a continued negotiation of an international fisheries agreement.

Cullis-Suzuki's research was focused on the effectiveness of what are known as regional fisheries management organizations—international institutions established to manage and conserve fish stocks on the high seas. The results of her research indicated that the organizations were not performing particularly well, neither in terms of the measures they said they were taking nor in regard to the state of the fish stocks they were set up to manage.

In a summary of her experience not long after the event, Cullis-Suzuki wrote, "If someone had told me this is where I'd end up, I would never have believed them. Yet here I was, just six months after completing my degree, about to address a room full of delegates in suits seated behind country name plates, and wondering what the heck I was doing."[19]

She goes on to describe her awe at being at the UN—a place that is certainly filled with mystique and excitement for most visitors. However, that wonder was quickly eclipsed when the event began. As one of four speakers on the panel, Cullis-Suzuki was first in line and says that as soon as she finished her talk, a diplomat raised his hand to tell her "just how wrong" she was. His comments, and those from others in the room, felt relentless. "All my experience had been in academic situations. I was ready for quantitative questions, questions about my data, but I was not prepared for these very emotional questions."

She says, "Of course I understood they had to defend their organizations, but it saddened me to hear them pick out and argue the mundane details of my study and painstakingly ignore the big picture." These are sentiments that many scientists who have waded into the science communication and policy worlds can probably relate to.

Understandably, she didn't feel great when the event was over—she felt guilty because she had upset people. Simultaneously, however, she herself was also upset because many of the comments and questions reflected an unwillingness to truly engage with her research results. "At that moment, I understood why people, especially scientists, don't speak out," she recalls.

The event had been stressful, and as the daughter of geneticist and environmental activist David Suzuki and activist Tara Cullis, as well as sister of Severn Cullis-Suzuki, who spoke before the 1992 UN Earth Summit at age twelve, she was perhaps more prepared than most to speak truth to power. However, there is little advice that can substitute for the actual experience.

Once she was back home in Vancouver and had time to reflect, Cullis-Suzuki realized how far her research results had traveled because of the UN event. "It occurred to me that had my language been anything less than strong, my speech any less direct, my conclusions less severe, the audience at the conference would surely have been half the size."

Even a decade later, with a doctoral degree and more experience under her belt, she's not sure she would have done anything differently. "I wasn't fully prepared in terms of understanding who I'd be talking to when I walked into that room. On the one hand, it's always important to know who your audience is. But at the same time—and I go back and forth with this—had I understood who the audience was composed of, I would have used different language. And that would have changed what I said, which I'm not sure would have been the right thing either."

A Relationship Challenge

As the many narratives here demonstrate, science communicators and the communities they work with are all exposed to what can be sad, scary, frustrating, and confusing information. Taking emotions and conflict seriously means that practitioners are facing much more than a communication challenge—they are facing a *relationship* challenge.

This is clear to Nadine Lymn, a science communication practitioner for going on three decades, based on the evolution of her own career. "I didn't necessarily start out focused on connecting people, but looking back, it's the thing I'm most proud of in my career." Like many others who were involved in science communication before the field was so named, she began working at the boundary between scientists and journalists and was quickly hooked.

As a graduate student in environmental studies at the University of Wisconsin–Madison in the late 1980s, Lymn was working with a local environmental radio program. She discovered she loved talking to researchers about their work and finding just the right way to frame a story. "That job gave me a thrill like no other, to the point that my graduate work started to suffer. My advisor eventually had to pointedly ask that I spend more time on my thesis and less time in the studio. That was a wake-up call that helped me realize I was more interested in the art of making research relevant to people's lives than I was in doing the research myself."

However, figuring out how to proceed after that realization was complicated. "Back then, there were two ways to go when you had dual interests in both science and communication. One was to become a reporter, and the other was to go into what was generally referred to as public affairs. I found a job that I loved in public affairs with a professional scientific society, the Ecological Society of America, where I ultimately spent over twenty years."

In that job, one of Lymn's first assignments was running a press room for the annual meeting, which that year happened to be in Madison. Because she had connections with local reporters, many more of them showed up to the conference than she had anticipated. "A television crew showed up on-site with their cameras, and I could not find anybody willing to talk to them on camera." Lymn says the inability to find scientists willing to talk with journalists "was a huge turning point for me in terms of understanding that we needed to start preparing our members to engage more broadly if that was their goal."

At first, says Lymn, there was resistance to supporting her public affairs position because there was a sense that journalists would find scientists through their peer-reviewed papers or other mechanisms. But, she says, things shifted over time, and "members clamored for public affairs offices, in our society and others, to assist them."

Lymn responded, and her workshops went from training scientists to talk with the media to how to converse with legislators and eventually to things like improvisation to encourage more fluid interactions with local agencies and community groups. She embraced her behind-the-scenes role, connecting scientists with one another as well as with journalists and policy-makers, and over time saw more and more scientists interested in working at smaller scales.

In her current role as the communications director for the National Science Board of the National Science Foundation, Lymn continues to focus on connecting scientists. "The scope and scale of my efforts have changed, but I still try to focus on what works best to ensure that the board maintains healthy relationships with Congress and the administration. Over time, I've come to see this relationship building and maintenance as the most complex and fundamental work there is when it comes to communicating science."

While to some it may sound a bit touchy-feely, when practiced seriously, relational approaches are among the most transformative and rewarding efforts that can be undertaken. They call on significant subjective skills, demand emotionally intelligent interactions that account for power, and recognize that while disciplinary knowledge is important, it isn't enough.

Relational approaches ask practitioners to see how they relate to one another as a fundamental piece of the work they do because what emerges from relationships is greater than the sum of its parts. They encourage a breakdown of the objective authority paradigm that science often relies on. And they stress that while the practitioner can function as an expert, they are "in" the work, too, which gives a way to recognize and deal with power dynamics.

Luckily, scientists and science communicators do not need to reinvent the wheel. Relational or relationship-centered practices are common in other fields including law and medicine, as well as having long-standing cultural roots. The following section explores relational processes that can help to guide contemporary science communication efforts on emotional and contentious issues.

Takeaways

- Emotions are present in science communication—and indeed all science—work, whether we like it or not. Ignoring them is to our peril.
- Emotions show up in all aspects of science communication. They aren't just ours, or just other people's, and the "collective stew" of emotions is crucial to understand.
- It is critical to question whose emotions are being centered in any discussion about them.

- Conflict is prominent in the emotional landscape of science communication.
- Taking emotion and conflict seriously in science communication means taking relating with others about them seriously.

Questions for Exploration
- What issues did reading this chapter raise for you?
- What are some of the ways emotion—your own and other people's—shows up in your work?
- What are some of the ways conflict—your own or other people's—show up in your work?
- Who affects or is affected by your work?
- Whom are you accountable to in your science communication efforts?

PART II

The Tools of Science Communication

CHAPTER 4

Relating

A TENURED PROFESSOR FROM A MAJOR RESEARCH UNIVERSITY stood in front of a room of people gathered to learn more about her recent work. Her research was on a topic that affected them directly and would ultimately impact how federal, state, and local governments regulated activities in their community. After a well-crafted and well-executed presentation on her research and results, the questions from community members rolled in. The first few were straightforward enough—they were about her methods, about the meaning of several data points, and about her proposed solutions.

But then, as can happen in community settings, the mood in the room began to shift. A woman stepped forward to ask a question that led the conversation in a contentious direction about local politics; another person stood in the back of the room, appearing angry, although it was unclear if the reason was related to the presentation or the question. People moved uncomfortably in their seats, and the scientist herself became uncomfortable. What had started out to be a familiar presentation turned into something nerve-racking for all involved.

For scientists and science communicators working in the kind of community-engaged settings described in previous chapters—academics at universities, government researchers, experts at nonprofits, and others—this is not an unfamiliar scenario. Perhaps it's a graduate student giving an initial presentation to an agricultural commodity group, a postdoc working with a watershed group, or a practitioner collaborating with a nonprofit.

There comes a critical moment when all the preparation in the world aimed at giving a good talk fails. The time spent framing and messaging and developing easy-to-read graphs falls away as somebody asks a perfectly human question. It is in those moments that the need to perform well gives way to the need to relate well.

If the early iterations of science communication focused on performance and providing information, later iterations are coming to center on relationship and knowledge building. Frequently missing, however, are related discussions about power and accountability. Research and practice from many other fields—including medicine, law, and psychotherapy, alongside long-standing cultural viewpoints and practices—have much to offer science communication practitioners today.[1]

A Relational Turn in Science Communication?

In 2010, science and technology studies expert Sheila Jasanoff concluded an article in *Science* by observing that the scientific community "has demonstrated that it can learn and change in its methods of representing science to scientists. That ingenuity should now be directed toward building relationships of trust and respect with the global citizens whose future climate science has undertaken to predict and reshape."[2] This bold and graceful statement signaled a shift in climate-related work by concluding that the scientific basis of climate change was well established; it was the now human-to-human element of the work that needed attention. Crucially, Jasanoff emphasized not just the centrality of human relationships but also accountability "between science and its publics."[3]

Although Jasonoff was focused on climate change, her statement applies broadly. Climate change is only one subject for which a science-based understanding has led to changes in global, national, and local policy. What she made clear was that in order to take action on highly complex matters infused with local and global uncertainties, many diverse people and communities would need to work together, or at least side by side, with accountability. A decade later, there are many organizations ranging from the Sunrise Movement to Extinction Rebellion leading climate work alongside artists and writers who are able to articulate ideas and stir emotions in a way science often cannot.

Accepting the kind of framework that Jasanoff laid out—one of relationship and accountability—situates many kinds of high-stakes science topics in a complex web of human relationships, inviting a sea change in science communication efforts. To expand from a focus on generating scientific knowledge into supporting long-term, transformative social change that improves people's lives—as determined by them—it will no longer be enough to rely purely on information, expertise, authority, and compliance. Instead, effort must shift to working directly in relationship with others, based on their knowledge and needs, from a stance of mutuality, reciprocity, and respect.

Defining Relationality
While there are many ways to look at relationships—from evolutionary perspectives to social group theory—the concept of "relationality" is one that offers a great deal of promise and is vastly underexplored in the sciences.[4] In a general sense, *relational* refers simply to connections, or relationships, between two or more people or things. For example, we might think about the relationships between people in a place and how they relate to that place.

Relationality as it is discussed today, however, has moved well beyond colloquial definitions and is itself a focus of scholarship and practice across multiple fields. Medical,[5] legal,[6] interreligious studies,[7] and geography[8] are just some of the fields to take up relational theory and practice. Business schools, professional programs, and sustainability education have all begun using relational methods and ideas. In many fields this focus is relatively new. Yet one place relational work has been deeply examined is in psychotherapy, where there is a rich history of theory and practice on the topic.[9]

Instructively, relationality in psychotherapy arose after decades of practitioners and researchers viewing the relationship between patient and analyst as one primarily about the problems of the patient, casting the practitioner as an objective and impartial expert. This view, of course, was eventually challenged from many angles including feminist theory and direct practitioner experience.[10] Based in part on these provocations, a new form of practice emerged that focused on what has been called a "two-person therapy" in recognition of the importance of the relationship itself, regardless of treatment modality.[11]

This is a necessarily simplified view of relational work in a single field. However, the emphasis on both participants being "in the work"—that is, the practitioner is not seen as the ultimate authority but instead understands they have much to learn from those they are working with—is an important component of relational work that has strong implications for science communication. Too frequently, communicators are placed or place themselves in expert roles that do not acknowledge the understanding and wisdom of "nonexperts."

The kind of relational work described here should not be confused with more informal understandings of relationships that can concentrate power for certain individuals or groups in nontransparent ways. That is, relational work should not be aimed at replicating an exclusive "good old boys" club with different members. Instead, it is crucial to consider who we are and are not in relationship with, how power is distributed within these relationships and affects those outside of them, and how to develop partnerships with equity and accountability.

Beyond professionalized outlooks on relationality, many cultural and religious traditions, as well as associated scholarship and practice, have integral relational practices, perspectives, and groundings. For example, "right relationship" is an overarching principle in some forms of Buddhism.[12] Relational frameworks emerge from Black feminist thought[13] and include feminist standpoint theory.[14] In addition, as described by scholars including Robin Wall Kimmerer,[15] Leanne Betasamosake Simpson,[16] Kim TallBear,[17] and Zoe Todd,[18] for many Indigenous people, relationality is fundamental and refers to relationships not only between people but with the nonhuman world as well.

As an example, Melanie Yazzie, an assistant professor at the University of New Mexico and cofounder of The Red Nation, coined the phrase *radical politics of relationality* in her work on resource extraction on the Navajo Nation.[19] Yazzie says, "I noticed that with *water is life*, the word *life* in particular was about countering the politics and reality of death that resource extraction has brought to native communities, including my own. I call these 'relations of extraction'—one-way relationships where resources are extracted from native lands with no benefit to native peoples." She continues, "That leads us into this radical politics of relationality based on Indigenous understanding of kinship, of relatives, of being in good relations of reciprocity. Those kinds of relations are about protection, about compassion and equality, about a power dynamic that isn't based on extraction and exploitation. It is about mutual respect and simply being a good relative."

In addition, Dominique David-Chavez, a postdoctoral fellow with Colorado State University and the University of Arizona's Native Nations Institute, is working to honor and support Indigenous approaches and contributions in the sciences.[20] Her work describes Indigenous relational frameworks as centering values including humility, integrity, respect, and reciprocity and working outward to develop research questions guided by those values.[21] In reviewing 125 climate research projects that involved Indigenous knowledge, David-Chavez and her colleague found that close to 90 percent of those projects practiced an extractive form of research "in which outside researchers use Indigenous knowledge systems with minimal participation or decision-making authority from communities who hold them."[22] Results like these underscore the accountability effort that Jasonoff articulated.

Frameworks for engaging with Indigenous scholars and communities include the concept of "ethical space" as outlined by Willie Ermine, a faculty member of the First Nations University of Canada. He describes ethical space as "formed when two societies, with disparate worldviews, are poised to engage each other."[23] Ermine's approach emphasizes equality and respect between Indigenous and non-Indigenous people's knowledge systems.

All of the many relational worldviews, practices, and related scholarship have unique aspects, yet together they demonstrate the widespread idea that relationship is simply fundamental in much of life, both human and nonhuman. Recognizing the commonalities between relational worldviews across cultures and traditions—understanding the world relationally versus transactionally—opens the possibility of working together across subjective and different ways of knowing. For example, better understanding the role of values in science communication might mean supporting and intersecting with the climate work that scholars like David-Chavez are doing in more meaningful, and accountable, ways.

There is much to be gained by looking at science communication, particularly on emotional and contentious issues, through a relational lens. As we have seen in the relational psychotherapy context, emphasizing the emergent and transformative properties of the relationship between practitioner and client has fundamentally changed the practice. Rather than being a by-product, the relationship itself is a guiding force in the work. Many would argue that this shift has been revolutionary because it focuses on facilitating change from within an individual rather than exerting it by an outside expert and has fundamentally changed the practice itself.

Thinking relationally means centering the role of relationships in science communication, not viewing them as peripheral or separate. This possibility of profound change—of transformation—has direct implications on science communication practice and could lead to a dramatic shift away from the present focus on expert-driven, coercive behavioral change.

A relational approach is revolutionary in that it asks that we take a giant leap into trusting not just scientific information but also each other. Fortunately, there are many examples of people already working relationally, in both the sciences and other fields, who can help to guide the way.

The Heart of Medicine

"Her heart never got a chance to heal," says Rebekah Fenton, MD, a pediatrician and Adolescent Fellow trainee at Lurie Children's Hospital of Chicago. She recalls the story of a teenager who had a heart transplant and was not sticking to her treatment plan, leaving her medical team frustrated and worried.

"My role was to go in and listen to her," Fenton says. "I discovered that she wasn't following her treatment plan, not because she didn't know what she 'should' do but because she was facing so many other challenges." As she sat with the teen that day, Fenton learned that the young woman had experienced multiple traumas that included the death of her father shortly after her own surgery and the marriage of her mother to a man whom no one in the family approved of.

"Just being able to sit and listen to what this teen was facing made me think about what it means when we as doctors get too concerned about things like how closely a patient is following the advice that we prescribe.

It's important, of course, but we can lose track of what we are here to do, which is to support people in healing."

When Fenton first began to work with what the medical community calls patient adherence, it seemed to her, as it does to many others, primarily about education. However, she soon realized patients experience many other obstacles to adhering to doctors' advice that have nothing to do with knowledge about what they should do. She says, "With chronic diseases, young people might need to take many different medications or have restrictions that their peers don't have." Fenton says connecting with patients and asking about their individual barriers and strengths is the thing that actually helps.

"There are so many areas where just relating and listening to patients is the answer. And it's not just the patients. I have to think in terms of the community of people around our patients as well." She often works with parents whose infants might slow, or even stop, eating solids when they have a mild illness and therefore not gain as much weight as growth charts indicate they should be.

"As a doctor, that feels perfectly normal because I have the experience and knowledge that indicates that a vast majority of the time, babies will be back on track with their feeding within a few days." Yet it can bring up immense feelings of guilt for parents. She has had to learn over time to not dismiss parents' concerns and to see that for them, it can be scary and heartbreaking, and she needs to validate those feelings before taking further action. "Part of that means being clear with myself about how I feel going into any given situation so that I can center their needs in what can be hard moments. And, of course, it helps to have supportive coworkers in these stressful situations."

Fenton says that taking time to connect with patients and their loved ones, as well as with herself and her colleagues, can be time-consuming, yet she's sure that it makes her a more effective doctor. Her experiences demonstrate the kinds of skills useful in clinical and practice-based work.

What a Relational Approach Offers

In today's communication environment, we are all exposed on a daily basis to scary and anxiety-inducing information about the world. From climate change to pandemics to artificial intelligence, emotion and controversy abound. The attendant grief and anger can be overwhelming. Learning to effectively interact in this emotionally laden environment calls on significant subjective skills, emotional intelligence, and the capacity to deal with uncertainty.

Simultaneously, many disciplines are shifting from an emphasis on the role of conscious cognition to an acceptance of the vital role of largely unconscious emotion. For example, in psychoanalysis, the words spoken between the analyst and patient might represent only a small part of the communication taking place; the nonverbal, unconscious aspects of communication are easily as important in the growth and transformation that can take place.[24]

This basic recognition—that the emotional, affective world plays a much larger role in our lives than we realize—has implications for how we think about scientific matters, particularly in how these issues are generally discussed in the realm of the intellect rather than the emotional. Cognitive and behavioral psychology have added a great deal of understanding to the science communication conversation.[25] Still, a dominant focus on cognition and behavior can be limiting if we accept that people can't always access the ways that feelings of fear, anxiety, and grief might impact their actions.[26] Understanding that people relate to each other on a level that may have little to do with what is being verbally expressed has the potential to alter communication methods and priorities.

Relational practice can be complex because it places the "expert" directly into the work, not standing outside it, encouraging a breakdown of the objective authority paradigm. It may resonate first with students—many of whom already understand that the contexts in which they will be working are full of emotion and conflict—who may see relational techniques as particularly helpful. Or perhaps it might catch the attention of those with long careers behind them, who may have experienced frustration with the accepted principles of scientific engagement on controversial issues that led to less-than-hoped-for outcomes.

Focusing on information dissemination—operating from the information deficit framework discussed in chapter 1—has in many cases been counterproductive for contentious scientific issues.[27] Simply supplying people with information about the negative effects of climate change or a multitude of other scientific topics is not necessarily working to create change. It is increasingly apparent that multipathway communication is called for. This kind of communication is necessarily a relational process.

The relational process involves developing an ability to listen deeply rather than focusing primarily on speaking or writing more effectively. It involves working with the emotions that inevitably arise when one truly listens to another person with curiosity and a willingness to change one's own views.[28] This is a different style of listening than is emphasized in the classroom, where students are more likely to learn to listen with an ear toward incorporating information developed by experts or defending their ideas. Instead, learning to listen actively and deeply in order to connect, empathize, and support rather than antagonize and change others' behaviors is an important aspect of continuing to evolve relational practice in science communication.

Many scientific issues—climate change, vaccines, GMOs—are increasingly taking on the characteristics of seemingly intractable conflicts and may remain unresolved despite good faith efforts.[29] To this day, the conflict around these subjects has been addressed as a science communication or behavior change challenge. Science communicators tend to focus on employing tools such as messaging, framing, and rewards for favored behaviors rather than more relational perspectives that include conflict as a key process that can be a driver of change.[30]

Explicitly and directly including relationships in science communication will require a significant effort because they have the potential to upend some of the most widely utilized approaches to science communication—such as reliance on scientific authority and compelling changes in behavior. It is a difficult proposition because working relationally places the practitioner directly into the process. No longer viewing oneself as outside the communities they study or work or teach in, practitioners as individuals, along with their reactions and emotions, become a pivotal part of ethical practice.

Learning Relational Practice from Other Fields

Fortunately, science communicators can benefit from a wealth of practices in a variety of fields. Some training and professional development programs for fields such as medicine, law, and psychotherapy include the value of relational work explicitly. This is in part because the professions involve direct person-to-person interaction in the form of lawyer-and-client or doctor-and-patient relationship situations, as well as situations where strong emotions such as grief[31] and anxiety[32] are present for both practitioners and those they interact with. The relational approaches developed in these fields offer important lessons for science communication and engagement.

Relationship in Medicine

As early as 1969, the concept of "patient-centered care" appeared in the medical literature.[33] However, it was not until 1992 that a task force noted that while biomedical science will always be at the core of academic medicine, patient relationships are also important.[34] A new literature emerged as researchers and analysts sought to understand and communicate the implications for incorporating a relational component into medical training and practice. Key concepts included the humanity of the practitioner, emotion as an integral aspect of practice, doctor-patient relationships as mutual, and practitioners as belonging to a larger community that includes other practitioners and communities.[35]

However, even with this explicit recognition of the importance of relationships in the medical profession, and despite the fact that at least some of the effort was driven by practical, insurance-related needs, many would observe that it has been a long path toward beginning to incorporate relational processes in practice. For example, Sharon Dobie describes the thorniness of a "hidden curriculum" of medical training: what leads many students to pursue a medical education can be lost with the pressures of training.[36]

"There can be resistance from students, faculty, and administrators," says Juliet McMullin, a professor of anthropology who has spent many years codeveloping the medical humanities program at the University of California, Riverside. Today, much of what can be thought of as relationship-centered medicine resides under the banner of medical humanities. McMullin says that in any given year, they could have as few as four students sign up to take a medical humanities course. Based on her conversations with colleagues at other universities, this is not unusual.

Fenton echoes these sentiments, saying that emotional intelligence and listening and all the things that fall under a relational umbrella "are still a sideline issue in medical school." One positive development is the kind of humanistic perspective that McMullin teaches, which Fenton says "allows us to talk about how to be human again" because medical training can serve to train the humanity out of people through the hidden curriculum.

It's a constant challenge, though, Fenton says as she recalls a paper that examined whether it was better for a doctor to be "nice or well trained." She says even that framing is telling because it still assumes that a person cannot be both. Fenton says she got that message early on in medical school, which resulted in a feeling that she wasn't as smart as other people because she wasn't a "fact rambler" and was more of a "feeling type." She second-guessed herself but through the training process learned that emotional intelligence was a real skill and regained her confidence.

Fenton doesn't feel alone in her humanistic outlook today. In fact, she sees that there is more appreciation for the value of emotional intelligence than there might have been in the past. McMullin agrees, noting that medical humanities programs are growing across the United States, despite the resistance.

Relationship in Law

There are several important and connected threads of relational legal work—including therapeutic jurisprudence, relationship-centered law, and restorative justice. Therapeutic jurisprudence prioritizes the psychological and emotional health of the people who come into contact with the law and legal professionals as one of many important considerations in legal practice.[37] The complementary field of relationship-centered law focuses on developing a more humanistic and service-based approach.[38] As in many fields, legal education has tended to focus on the intellectual, but a shift to a more relational style includes a focus on the development of affective, or emotional, skills.

There is overlap between medical and legal relational techniques that serve to reinforce crucial methods. Both medical and legal approaches advocate for linking cognitive and affective skills. Training and practice for science communicators could be enhanced using similar training methods to complement existing efforts. Hand in hand, and similar to developments in the medical field, contemplative practices have also been gaining a foothold in legal training and work.[39]

Gail Silverstein says straightforwardly, "I think lawyering is essentially a relational profession." In her role as a clinical law professor at the University of California's Hastings College of the Law, Silverstein works with law students to infuse a relational orientation into their training. After spending several years working with a community-based law clinic as a staff attorney and clinical supervisor in the HIV/AIDS unit where she specialized in legal services for women and children, Silverstein understands the dissonance between training and practice quite well.

"Although relationships are fundamental in law practice, usually the only exposure that students get to relational skills is in the clinical classroom." When law students arrive in her classes, they are encouraged to work collaboratively and do what they can to make the space feel "safe"—that is, accommodating and supportive—for one another. "Law school is competitive and individualistic, so I have to set a different context for clinical work. That includes modeling a collaborative orientation myself and making the classroom reflective of the kind of relational work that we are encouraging. We can't just say it, we have to do it, starting in the classroom."

Like McMullin, she says that fitting practice skills into legal training is difficult when there is so much to learn, and the hidden curriculum creeps in during law school as well. "The students can definitely be resistant to relational work, and it's sometimes just scary and overwhelming for them, especially at first. Sometimes it takes accompanying a student to a one-on-one visit with a client to help them to feel more comfortable talking in person as opposed to just texting."

Silverstein also encourages students to think about different models of lawyering, as well as the context in which they will be working. "It is valuable for students to understand how many different ways there are to be a lawyer, and that can help guide how they want to relate to clients. We talk a lot about the client-centered approach, but it may not always be appropriate. And practicing corporate law is generally different from social justice law. I want students to have a good sense of not just who they are as lawyers but also how they want to work with their clients."

Relationship in Psychotherapy

As mentioned earlier in this chapter, relational work has been well described in psychotherapeutic literature and deeply explored in practice. "I think there are many layers to therapeutic relationships, from building rapport to establishing an alliance. I try to think of my therapeutic relationships as standing with a client through uncertainty," explains Leslie Davenport, author of the book *Emotional Resiliency in the Era of Climate Change*[40] and practicing psychotherapist. "I don't have the answers, but I do have this therapeutic lens that I can use to support what is emergent for them and in our work together." Davenport notes that being trained as a marriage and family therapist also means being attuned to understanding relationships within the context of a larger system that goes beyond the individual.

Working relationally is also key for Theopia Jackson, who says simply that "relationship is fundamental in psychotherapy." Jackson is a clinical psychologist and program chair at Saybrook University with a long history of providing child, adolescent, and family therapy services who has also begun to speak about climate change in Black communities. "When I step back and think about what that means after so many years of practice, it comes up in many different ways. For example, I can think of my practice as a service model, where I frame my efforts around being in service of others. That means that I can't come at it from a place of being an expert with all the answers."

This insistence on allowing the wisdom of others to come through is something that can take practice, but Jackson has concrete suggestions for dealing with the urge to provide answers for others. "Instead, maybe I can come at it from a place of asking, 'What do you need? Here's what I can offer. Is this helpful? If so, how do I make it accessible to you? If not, what do you need that I might be able to provide? How can we look at this together? How will you know if this is helpful?' Those can become guiding questions for working relationally."

Relating as a Science Communication Goal

Reyhaneh Maktoufi is a science communication researcher, trainer, and practitioner, currently serving as a Rita Allen Foundation Civic Science Fellow at *Nova* and as a producer at The Story Collider. Even before beginning an official first career as a physiotherapist in Iran, she volunteered in hospitals, working with children. There, she says, she started to learn communication skills. "I was learning how to talk to people, how to ask for permission, particularly with children, how to be open, how to listen, and how to sit with grief."

After moving to the United States and embarking on a doctoral degree in science communication with a focus on curiosity as a communication goal that can lead to deeper connections, Maktoufi says that the lessons she learned early in life applied in her new research and practice as well. "When I started my research, I realized that so many of the challenges in science communication were similar to those in the health field. There was a big focus on persuading people to do things, as opposed to connecting with them. There was a big focus on changing the perception that scientists were untrustworthy, as opposed to how scientists can actually be more trustworthy."

To address this, she started to teach workshops focused on empathetic science communication for more advanced students. One technique she uses is to ask students to act out different personas—one might be asked to take a stance as someone who is skeptical about climate change, while the other is someone who is not. She puts bounds on the exercise, telling the climate skeptic that they are also well educated and civic minded, to ensure that the exercise does not involve simple stereotypes. Her hope is that students walk away with a deeper understanding of the value of connecting with others and not simply persuading.

She has seen this in her own work. While observing scientist and visitor interactions at the Adler Planetarium in Chicago as a doctoral student, Maktoufi says that she watched a school group begin a conversation with a scientist about the origin and evolution of the earth. She says, "Pretty quickly, the teacher objected and said their students didn't believe the earth was that old. I did see a quick moment of adjustment on behalf of the scientist, but luckily she was able to pivot to talking about galaxies—she rolled with it. Afterward, the teacher warmly said thank you and expressed appreciation for how the scientist had handled the comment."

Maktoufi notes that had the scientist been offended, it could have been a very different interaction. "Had she had a goal to immediately persuade this group of the age of the earth, her communication effort would have been seen as a failure. But if her objective was to inspire curiosity and open a door, then I would say it was a successful interaction."

Besides, she asks, "What was the alternative? For the group to leave? To hate science and scientists? I think that conversation at least opens the door to further conversations, makes the planetarium a more welcoming environment, and shows that scientists are willing to connect and be curious together. You just lose a lot of opportunities when the immediate goal is to persuade instead of connect."

Putting a Relational Approach into Practice

One of the primary questions that surfaces when discussing relational approaches to science communication and engagement is how to put the concept into practice. The narratives shared in this chapter are just a handful of many that can be helpful in thinking about the role of relationships in the evolution of science communication and engagement practice.

In many ways, the easiest place to start is with increasing the relational skill and capacity of individuals. Other professional fields offer an initial roadmap. Contemplative practices are well established in the medical field and others as tools for helping individuals and groups to navigate difficult emotional terrain.[41] Existing practices can be adapted to complement current science communication training efforts[42] and in some cases are being performed already, as Maktoufi's empathy workshop indicates. Other key skills include cultural humility and accountability, the ability to contextualize, tolerance to emotional vulnerability, and self-awareness, which can be practiced through activities like listening, self-reflection, and journaling.[43]

There is a large body of professional development from various fields that could be used in these areas. Working with conflict can be addressed initially by professional development in facilitation work or courses like Crucial Conversations. There are also opportunities to engage in learning about emotional intelligence and trauma-informed practices in most fields these days. Training from institutions like Stanford's Center for Compassion and Altruism Research, the University of California, Berkeley's Greater Good Science Center, and the University of Virginia's Contemplative Sciences Center is available. Creative approaches to science communication can be facilitated through collaborative, interdisciplinary efforts with individuals and organizations already doing this work in different contexts.

These tools offer only the beginnings of relationship-centered science communication and are geared toward connections between individuals. Although relational practice theoretically encompasses working with power, in reality, this can be much harder to do. Working relationally might simply feel better to practitioners, but it doesn't always mean that it is truly getting at the accountability that Jasanoff emphasized.

Yanna Lambrinidou, an anthropologist who specializes in environmental health, says her "experience with efforts to cultivate humane health professionals is that, as important as they are, they may not always question the presumed superiority of biomedicine over other types of healing systems. Similarly, they may not always challenge the intrinsic power of health professionals over patients, no matter how relational and empathic these professionals might be." She adds, "In contested spaces of health and illness, a professional's failure to recognize that unfamiliar patient experiences, ontologies, and epistemologies have the capacity to expand, improve, and even correct conventional biomedical practice can result in serious harm despite the professional's best intentions."

Lambrinidou stresses that the humanization of professionals cannot come at the expense of patients. "At the end of the day, I think we want to make sure that projects of 'humanization' do not end up stroking the liberal sensibilities of health professionals more than equipping them to practice in a manner that patients experience as effective, appropriate, and just." In other words, she says, "'Humanization' rather than making the 'biomedical pill' kinder, gentler, and easier to swallow must produce serious conversations with patients about whether this pill is the proper response in the first place and who should get to decide."

"In ecology, for example, 'Traditional Ecological Knowledge' can be used as a bargaining chip, a tool for individual scientists to humanize themselves by claiming that they support 'decolonization,' an effort that aims to dismantle colonial structures," says Sara Kahanamoku from chapter 1. "Yet these projects are often highly exploitative because scientists determine which questions to ask and which priorities to pursue, rarely stopping to consider whether they should have power over communities to make these decisions in the first place. This is 'decolonization' in name only. It doesn't actually give power to Indigenous people, and in attempting to humanize scientists, we are dehumanizing other people and communities in the process."

As relational science communication practice matures, there are many nuances that will continue to emerge, and they will need infrastructure for support. Even within the framework of strong institutional support, there is a need for "reflexivity"[44]—consistent reexamination of how knowledge is produced and particularly how the producer shapes that knowledge—and adaptation to ensure that relational approaches are equitable and ethical.

Reflecting on Relational Work

Relational work in science and science communication may seem new, but it is not. One of the most enduring models comes from the US Cooperative Extension System, which was established over one hundred years ago. As described in chapter 1, cooperative extension is integrated into the land-grant university system. The cooperative extension model was in part based on the early recognition that scientific research should be developed and communicated with the people affected by its results.

In contrast with traditional academic research, which is commonly initiated in response to calls for proposals from funding agencies and may have limited support for broader impacts, extension projects are typically generated through interactions between county and campus-based academics and community members. Unfortunately, extension practice has remained quite separate from typical science communication practice. This is in part because extension already combines research and communication in a way that makes them difficult to parse.

One thing that stands out about cooperative extension is the commitment to long-term relationships with communities, whether they be farmers or youth or government officials. Relationship building is a key part of extension work and something that individuals who seek out extension positions continue to hone throughout the course of their careers.

Clare Gupta, a cooperative extension specialist at the University of California, Davis, has done research on the value of the cooperative extension system. With collaborators David Campbell and Alexandra Cole-Weiss, Gupta found that while cooperative extension administrators generally point to technical expertise and objective, science-based information as key strengths, relationship-based strengths may be overlooked. After compiling a set of case studies of extension project outcomes, they write that their research highlights a story "that embraces both technical and relational work and that communicates the ways in which, in concrete settings, we weave those types of work together."[45]

As an example, Michelle Leinfelder-Miles, a farm advisor with the University of California Cooperative Extension system, says that she practices "little *s*" science—research developed through conversations that directly addresses local needs—in an effort to develop trusting working relationships with growers in California's Sacramento–San Joaquin River Delta region. Although she comes from a sixth-generation farming family in the area, she laughs recalling that because her family farms in the same county but outside the Delta region, some farmers joke that she "came from a completely different world." She shares, "It was funny that I considered myself local, but I was not local enough! The 'little *s*' science that I do has been so valuable in developing rapport and becoming a trusted member of the community."

As you might imagine, this work invites continual reflection to ensure that relational work is equitably distributed. In science communication practice, a common piece of advice is that practitioners focus on their audiences. In extension, however, it can be complicated to ensure that "audiences" do not become *too* narrowly defined. Because relationships are so fundamental, extension professionals tend to feel protective of them. While helpful in some situations, in others, that safeguarding of relationships can function as gatekeeping.

Kripa Jagannathan, a climate change adaptation researcher at the Lawrence Berkeley National Labs, set out to understand how farmers and growers might use long-term climate change projections in such on-farm decisions as what to plant and when. She interviewed both farmers and farm advisors and found that in some cases, the advisors were more conservative than farmers in terms of thinking about the types of climate projection information that might be useful. However, through a conversational interview process that focused on deep listening and open-ended questioning, she worked with both the farmers and farm advisors to develop a shared sense of what long-term climate projections are and what they can provide. "It was a deeply relational process with emergent qualities. Had it been a survey, the results would have been quite different. Instead, it felt a bit like a therapy process," she says.

This process resonates with Mark Thorne, a cooperative extension specialist with the University of Hawai'i at Mānoa. Thorne spent many years early in his career avoiding talking about climate change because it felt too politically contentious to go into with the ranchers whom he deeply values working with. However, after a session at an extension climate change workshop that included a focus on relational communication, he had a change of heart and realized "as part of our responsibility as extension educators, we have to provide reliable, science-based assistance," and that includes broaching subjects that may feel controversial.

"The consequence of avoiding talking about the effects of climate change on agricultural production has too many negative consequences for the people I work with and for food security in general at the local, regional, and national levels," Thorne says. "I now see it as a professional and ethical obligation to talk about what can be difficult topics, and I have to trust that the long-term working relationships I have are strong enough to handle it."

While a relational approach may sound basic to some and complicated to others, in both cases, it is helpful to know that there is evidence that relational practices can indeed be taught and learned and that learning them is something that requires effort for many. Even basic exercises, such as reflecting on the mutual benefit of a relationship between a researcher or practitioner and community member, can help to increase individual relational capacity.[46] However, the true promise of relational work in science communication lies beyond interpersonal relationships. In an ideal world, relational work would lead to a reconsideration of whom we are working with, how, and to what to end, as well as who gets to determine success.

Relationship is an active, emergent space that can feel uncomfortable to people who are trained to reduce uncertainty. It truly calls on a different set of skills in the practitioner. The next few chapters will outline some of the most relevant and crucial of those skills—listening, working with conflict, and understanding trauma—for effective science communication in service of society.

Takeaways

- Relationship-centered communication offers a strong set of theories and practices from other fields that can inform science communication practice.
- It is important to understand the difference between relational work grounded in theory and more colloquial understandings of relationships, which can serve to re-create and reinforce existing power structures.
- In many ways, a relational focus has been present in science communication for a long time, but it has been overlooked in the popular discourse of the twenty-first century.
- Science communicators can learn from and partner with doctors, lawyers, therapists, and Indigenous and other scholarship and cultural practices regarding relationship-centered communication approaches; there is no need to reinvent the wheel.
- Working relationally is a lifelong practice that invites continual reflexivity to ensure the work is being done ethically and with accountability.

Questions for Exploration

- What issues did reading this chapter raise for you?
- Whom are you already working with in relationship?

- What are some relational skills that you already have, and which ones would you like to practice more?
- What are some ways to ensure you are not contributing to unequal power dynamics in your work?
- Whom does your science communication practice serve, and why?

CHAPTER 5

Listening

LISTENING IS A VITAL COMMUNICATION SKILL but one that has tended to be overlooked or treated superficially in science communication. When listening is discussed, it is almost exclusively in the domain of interpersonal communication, rarely mentioning power and accountability. That is, at least in large part, because the dominant science communication model has been focused on information dissemination with an implicit sense that information is authoritative on its own. The input of those beyond the scientific community has not always been particularly valued. However, with newer models of relational communication, the question of what it means to listen well is more pertinent than ever.

Danielle N. Lee, a scientist and science communicator at South Illinois University Edwardsville, shared a powerful story about the role of listening in her own work in a keynote address at the 2019 Society for Advancement of Chicanos/Hispanics and Native Americans in Science (SACNAS) conference. Lee speaks of her experience traveling to Tanzania, Africa, to study pouched rats as a Black American researcher, the incongruity between how she thought the experience would be versus how it was, and the awakening she had there. To raucous applause, she says, "A stranger can't discover something in someone else's home."[1]

The course of events that led her to that realization started with listening. "I began to listen and I shifted my perspective. Listening is my most essential scientific skill. I listen with my ears, my eyes, my heart, my spirit. Listening to elders, the residents, the keepers, the Indigenous people of any place is the demonstration of the very first scientific step in the scientific method that we learn in school: observation. Listening is good science."[2]

A key question becomes, What does it mean to center listening in science communication practice? It's a field that has tended to focus almost entirely on output and not input. However, there are some inspiring examples to be found of leading with listening, valuable listening techniques to explore, and many ethical considerations related to what practitioners should do with what they hear.

Listening for What You Don't Know

"It's important to elevate community expertise to be on par with agency expertise. We have to get past the 'we know what they need' mode of resource management," says Mike Antos, senior integrated water management specialist with Stantec Consulting. "While that idea has a policy legacy in California, it took about ten years to be in the right position to undertake a project starting from that point."

Antos, a geographer by training, connected with Valerie Olson, an environmental anthropology professor at the University of California, Irvine, and Emily Brooks, then a postdoctoral scholar at the University of California, Irvine, and now a physical scientist at the US Geological Survey. Collaborating with several other organizations, the group developed a project aimed at gaining a better understanding of how communities think about and use their water and how Southern California water agencies can better serve them.

The group started with a needs assessment, but as Antos notes, "We reframed it as a 'strengths and needs assessment' to overcome the deficit model of thinking. And we centered listening as a key process. I've spent time as a 911 dispatcher, and I know how crucial listening well is." Brooks adds,[3] "The idea is that locals are the best experts on their own communities' needs and strengths. They live in and move through these places every day, and have experience and knowledge that isn't accessible to outside experts like water planners."

An important goal of the project was to find out how water agencies could best support people whom they didn't usually hear from, which includes the homeless and renters, because so much agency outreach is to homeowners. Brooks says that "water agencies have particular assumptions about what communities need. But when we talked to people in those communities, we heard a different set of priorities that included affordable housing, public parks, and roads. The challenge then became thinking about how water might fit into those priorities."

The team drew on ethnographic methods like open-ended interviews and listening sessions, hoping to develop a process that other water agencies and researchers could use. "We wanted water agencies to be able to better listen and respond to the communities they serve. We know a lot about certain water behaviors from certain people—for example, what incentivizes homeowners to take advantage of rebate programs—and almost nothing about others."

What they found through their listening process was a series of what they termed "connections and disconnections" between resource managers and various communities.[4] In other words, they found that the community is far from monolithic and does not experience water in the same way, and therefore the agencies need more specific interventions. Water managers might rely on water testing to determine that it is safe to drink and feel confident about the results. However, water users might rely on their experience of their drinking water tasting or smelling bad and decide that it is not worth risking their personal health for. These different perspectives can lead to a conflict between managers and users, even though they might connect around the basic need to have safe drinking water.

In the end, Olson says the listening process raised even more questions for water managers while uncovering key processes that need to be addressed in new ways. It illuminated, for instance, that language created barriers. The agency had been offering on-call translation services, but they were burdensome for users. Listening brought home the idea that a new effort, such as hiring multilingual staff, should be a priority.

Science communicators can learn from and partner with colleagues who have expertise in listening to undertake this kind of truly engaged work. However, any collaborations will also require co-learning for all involved, as even professionals who have listening at the core of their practice can still struggle to explicitly address listening.

What Does It Mean to Listen Well?
Listening makes conversation possible; otherwise, we may as well be speaking to a wall. In the context of science communication, listening has taken a back seat to speaking. That imbalance has implications for not only interpersonal relationships but also broader questions of whom science communication serves.

On an interpersonal level, listening is obviously valuable. However, what it means to listen well, and to what end, is rarely explicitly discussed. As children, many people are taught to listen by sitting still and giving their undivided attention to a speaker, frequently a teacher or a parent. That type of listening continues through most of the schooling process: first students are taught to absorb information verbatim, and then in traditional college settings, they are trained to listen so that they can critique.

These patterns can be hard to address because most people consider themselves good listeners, as a study by Jack Zenger and Joseph Folkman indicates.[5] "Chances are you think you're a good listener. People's appraisal of their listening ability is much like their assessment of their driving skills, in that the great bulk of adults think they're above average," the duo wrote in the *Harvard Business Review*. They found listening was critical for good leaders, which further sparked their curiosity about what it meant to listen well. Some of their results run counter to common wisdom. They found good listening was less about being a silent sounding board; instead, people enjoyed being asked insightful questions that indicated the listener was hearing what was being said. Their work suggests that there are many ways to think about good listening and that, as always, context matters.

There are many things that get in the way of good listening, including being distracted by devices or other people, having an agenda, or being short on time. There is also the fact that modern culture does not offer easy places for listening, so those spaces have to be intentionally created. And while there can be a sense that people must learn to listen "across the aisle," it is perhaps even more valuable to listen closely to those who are perceived to be similar because assumptions of similarity can test the foundations of even close friendships and partnerships. Some of the most difficult moments can come when people who are assumed to agree on most everything learn that's not the case.

Listening for What Is Really Being Said

"As a water commissioner, I sit behind a table with my two other commissioners and listen to members of my community. They come talk to us, tell us their concerns, ask us questions, and advocate for what they want. A lot of feelings come up in those interactions," says Erika Amir-Lin.

By day, Amir-Lin is a hydrogeologist with a large international environmental and engineering consulting firm in Massachusetts. She spends a couple evenings a month, however, serving her town. "Although I work on water professionally, my role there is to deal in cold, technical facts with a capital *F*. We are paid to not have feelings about things we are working on, to remove ourselves from disputes. But as a water commissioner, I am there to sit with people's feelings about water."

As an example, she says, three towns in her area share the water rights to a particular pond. When one town wanted to upgrade its water treatment facility, it caused a minor uproar in the other two. "Suddenly people's emotions were stirred up; they were saying, 'This water belongs to us.' But, the truth is, our town doesn't want to use the pond. We're not set up to treat the water. But people want us to say, 'That water is ours.' When they come talk to us, they're bringing that emotion. They passionately want us to validate their feelings, but we cannot, because we don't want that water, nor do we want to go to war with our neighbors over it."

Amir-Lin says that listening to people can be challenging:

> We have some frequent visitors that have latched on to ideas that are not scientifically correct. Sometimes it can be hard to figure out what people are even telling you. Some people have an agenda, and you have to factor that in. You have to listen in good faith but also not take everything at face value. You have to be your own drill sergeant because you're going to fall back into judgment and less generous habits toward people. You're managing simultaneous tracks in your mind, in addition to having people sitting a couple yards away, people you're going to see at the supermarket. I have to wear short sleeves even in winter because I sweat with the effort of being fully present!

The listening that Amir-Lin does is structured by a deliberative process that provides boundaries for how information is taken up, a framework that she says has pros and cons. "There are three of us, and over time we've learned to separate listening from responding to give us time to deliberate. It helps because I know there is time to digest before I act," she says. "I'm somebody who wants to make the world a better place right now. I've had to learn to rein in that impulse."

On the flip side, Amir-Lin says:

> While I appreciate the time for deliberation, there's a risk of becoming apolitical. We are not a partisan institution, but water is political and always has been. The board has tended to shy away from engaging with the messier human aspects of water management. When people say, "We're going to run out of water," what they mean is "We're going to have too many people in town." Which is a different conversation. It's actually a conversation about the vision for our town. But it becomes a nested series of conversations about water use. It's the conversation under the conversation that can be hard to get to.

Listening Techniques

Again, in the science communication and engagement realm, while there is more attention being paid to listening, it is not even close to matching the advice on how to talk. It's difficult even to imagine a course in something like "public listening" being offered, while public speaking training is ubiquitous. This is unfortunate; as the narratives shared in this chapter indicate, listening is complex, and there are many ways to think about and practice it, as well as ethical considerations associated with it.

This overview is by no means an exhaustive survey of listening techniques; instead, it is intended to provide some of the methods out there that might help as initial steps in taking up an intentional listening practice. From there, there are many resources for deeper dives on listening, including academic journals and articles, as well as many types of experiential learning opportunities.

Active listening is one of the most commonly encountered versions of intentional listening.[6] It goes by many names, including empathetic listening and reflected listening, and is used across many professional settings, from psychotherapy to nursing to crisis negotiation. As a brief search will reveal, there are many definitions of active listening, but most include three basic elements: nonverbal acknowledgment that one is listening, paraphrasing and repeating back to the speaker what is being heard, and questions that encourage further conversation. It is intended to be an accepting, nonjudgmental approach to listening with a goal of deepening trust and relationship.

There are times when active listening is just a little too, well, active. *Deep listening* is another way of thinking about listening. While it is also focused on nonjudgmental attention, it does not ask the listener to nod, repeat back what was heard, or ask further questions. Instead, it is about developing deep presence and allowing the speaker the time and space to both find and say what is on their mind, which by necessity may also involve periods of silence. That silence is what can be difficult for some people, particularly those trained to smooth over social situations with words.

In fact, allowing for silence can take extra effort in expert communities. Many scientists and communicators are trained to fill pauses in conversation with more words. Being able to respond quickly is important in many scenarios. However, learning that silence may at times be what others need most can help to justify the discomfort. Indeed, some of the hardest and most valuable conversations may invite a silence that can then allow for the emergence of things that might otherwise be glossed over or remain unsaid.

Companioning is yet another variation of listening that includes being able to sit in reflective silence with another person. The focus is on the listener practicing compassion and effective mirroring while listening. At heart it is about the listener working to bring out the wisdom in the person being listened to. That might mean understanding what the listener brings to the table first so that they can listen with focus, curiosity, and open-mindedness.

Embodied listening is yet another helpful practice. Many times, listening with the whole body can give information that might otherwise be hard to recognize. Maybe you notice when you are listening to certain people, your chest tightens, you stop breathing, your stomach clenches, you lean away. Other times, you might notice involuntarily taking a deep breath or leaning toward a speaker. All those bodily reactions offer valuable information. Unfortunately, many people are trained to ignore those kinds of messages starting at a young age. However, for a professional, a knot in the stomach, tension in the shoulders, or even butterflies might be signs to pay deeper attention to what is happening in any given situation.

Knowing when and how to listen has as much to do with *listening to the self* as listening to others. When listening becomes a deeply held value, making sure to also check in with one's own thoughts, feelings, and reactions, all while trying not to get too caught up in them, is a tricky proposition and certainly a lifelong discipline.

There are also times when *not listening* is appropriate, particularly when power dynamics make any other form of resistance dangerous or futile. Indeed, one of the shortcomings of contemporary discussions of listening is that they fail to address power and hierarchy related to who has to listen to whom. Sometimes there are harmful things being said or valuable voices being silenced, and the only self-protection that remains is to stop listening. Sometimes the listening process is not reciprocal, or feels extractive, and not listening is the only way to exercise agency. Not listening can be liberating, particularly when it is intentional.

Listening can be practiced alone, with another person, in a group, in a circle. It can be practiced indoors and takes on new dimensions outdoors. As Robin Wall Kimmerer wrote, "Listening in wild places, we are audience to conversations in a language not our own."[7] It can be done online. In fact, a big focus of social media managers is to use their platforms in "listening mode." And at times, taking breaks from all that listening, online and otherwise, can also be valuable.

Across all these forms of listening, one of the major themes is that listening should be a "positive" experience. That does not mean these are not difficult conversations; indeed, sometimes absorbing information that is counter to one's beliefs is difficult. It does not mean that conflict is swept under the rug or that power is ignored. But it does signal a shift away from how many scientists and science communicators are trained to listen, particularly in the graduate school process, which is with an eye to critique. Much of the research and practice on listening instead show that creating a hostile listening environment goes against the goal of building rapport and lasting reciprocal relationships.

Related Approaches
Sometimes the subtleties of listening may invite techniques for working with people so that they can open up. There are many methods for being in conversation that help get the dialogue started and keep it going. At the same time, it is crucial to pay attention to power dynamics and hierarchies to ensure that people are not manipulated into speaking when they could be harmed.

Open-ended questioning is a technique that supports continuing conversations by asking questions that cannot be answered with one-word responses like yes or no. Open-ended questions allow for information to emerge rather than to be forced out. In practice, this is often achieved by asking "what" and "how" questions rather than "why" or "when."

Motivational interviewing is another method that is not commonly applied in science communication but shows promise. There are different ways to think about motivational interviewing, but the basic framework comes from addiction treatment.[8] Many people are familiar with the concept of intervention, an effort focused on confronting addicts with the "facts" of their situation, how their behavior affected them and others, and breaking down denial. This technique, of course, turned out not to work all that well. As a result, motivational interviewing emerged as an alternative focused on working with people to support them in finding their own motivation for healing, which has been a big paradigm shift in addiction treatment and has applications for topics like climate change.

Nonviolent communication, or simply NVC, is a prevalent approach to communication.[9] It focuses on expressing how you and your communication partner are feeling without blame or criticism being offered or heard. Within that framework, communication moves from observations to feelings to needs to requests between communication partners.

An example might be that one partner says something like "When I remember that argument we had, I feel scared because I need space for a calmer discussion. Would you be willing to speak in a quieter voice?" It would then be up to the other partner to respond to the request. That partner might say in return, "When you hear a loud, argumentative voice, you feel scared, but you need to feel safe. Would you be willing to work on expanding your sense of safety?" In both cases, the goal is for the speaker not to blame or criticize and for the receiver not to hear blame or criticism.

Accessible Listening

Interpreted narrowly, listening can be perceived as something that is done solely with our ears. However, in its broad and embodied form, listening is something that everyone does, even if the styles vary. At the same time, ensuring that everyone can participate in listening and connect with others takes concrete action. Given that 15 percent of the adults in the United States report some trouble hearing,[10] this is important.

Gabi Serrato Marks, a doctoral candidate at the Massachusetts Institute of Technology who researches past rainfall changes in Mexico using cave stalagmites, is a science communication practitioner who addresses accessibility in STEM for people with disabilities. When it comes to listening, Serrato Marks says it's important to understand the different ways that people listen and process information. "Neurodiversity considerations are important. For example, doodling helps some people listen better. There are also cultural norms around listening that might vary—like whether making eye contact is considered polite or not."

Serrato Marks suggests that when it comes to practicing listening, flexibility and communication are invaluable. "During listening exercises, make sure people face each other, and make it easy for people to find a quieter space if they are in a room with background noise. When you're working with someone hard of hearing, speak at a normal pace and don't overenunciate. If you are using an interpreter, speak to the person you're talking to, not the interpreter—this is also true with spoken language interpretation."

Deborah Seiler is the Extension and Public Engagement Connection Center coordinator for the University of Illinois campus community—faculty and staff members and students—and the extension personnel in the state. With a master's in environmental behavior change, Seiler has a long history in community-engaged communication efforts. She is also deaf—she lost her hearing gradually, starting in her teens, and received cochlear implants in her midtwenties. Seiler notes that this is different from being Deaf, which in part means communicating primarily with sign languages such as American Sign Language (ASL).

"My perspective is only mine; I can't speak for all deaf people. Having trouble communicating is how I got interested in communication—I was obsessed with learning how to take a complicated message and make it simple. Because for me, if someone talks for a long time or in a roundabout way, listening is exhausting. If there is an event that is just talking without visual resources, it's too much. Adapting to so many online calls with COVID-19 has been hard for this reason. I have a keen interest in getting things across efficiently."

As an undergraduate, Seiler had a real-time ASL interpreter for all her classes. She notes that "listening is also impacted by interpretation, whether that's through our personal filters or literally across languages, including ASL. There's so much variation in hearing ability, including people who may need their partner sitting next to them to repeat what the speaker said. That means they're not getting the same message as everybody else, right? There have been plenty of times in my life I answered the wrong question, because I heard it wrong. Language and hearing ability is just one more way that we can experience the same situation in very different ways."

Seiler says that listening is fundamental to communicating. "I have to know who I'm interacting with and what they care about. If as science communicators we're representing research-based information that was arrived at with great effort and care, then we owe equal diligence to how we communicate, and that means understanding others," she says. "I have an obligation to make things as accessible as I can—like using closed captioning with videos that I make. I think it's important to realize that listening can happen in many ways, and whatever we can do to facilitate understanding, we should."

These listening techniques are just the beginning of thinking about how listening fits into science communication and engagement processes. The following project puts some of them into action.

Listening by Design

Cities throughout the world are facing ongoing infrastructure maintenance problems that are expensive and time-consuming to fix.[11] One of those is related to household wastes that are flushed out through drainpipes—whether that be the toilet or kitchen sink—leading to big problems in local sewer systems and beyond. City managers have found that one of the biggest challenges in addressing this waste is rendering visible the invisible connection between household wastewater and city infrastructure.

Working with the city of San Jose, California, as part of a unique collaboration between the city's environmental service and transportation departments, as well as its Public Art Program, Claire Napawan, a landscape architect and urban designer, and Brett Snyder, an architect and graphic designer, both at the University of California, Davis, engaged with communities to better understand how residents deal with their kitchen waste in particular.

Napawan says the city's initial request for proposals on dealing with waste—titled "Public Art Project for Environmental Awareness of Sewer System Impacts"—was a shock. It asked for exactly the kind of thing that she and Snyder do, which is work at the intersection of design, infrastructure, and community participation. "It was an amazing experience to see that the city recognized the value of communication, community engagement, and human-centered design from the beginning."

Napawan and Snyder took the participatory process seriously. After partnering with the city to identify waste hot spots, they then determined the best avenues to begin working with those communities. They quickly homed in on parents' groups associated with local public schools. The groups were already there to provide opportunities for community members to discuss concerns including education, health, and environmental stewardship.

From the beginning, the two were certain they did not want their community engagement to focus directly on the problem by assigning blame or instructing community members on best management practices. They instead set up meetings with parents' groups as conversations on "kitchens, communities, and the city" and invited participants to share a recipe during their first meeting. Napawan notes that the recipe share "opened the door to talking about the waste generated by these meals and how they are managed. Food became a great way to link the various concerns of the community with the sanitary sewer infrastructure."

During the project, Napawan was pregnant with her second child and conducted many of the early community workshops noticeably so, something that led to more connections. She says, "It became a way I could relate with other mothers. After my son was born, I continued to lead workshops—I would just bring him, pass him off to another mother, nurse during breaks, and have fantastic conversations about motherhood, as well as about sewers. This was such a rare and wonderful opportunity to feel like pregnancy and motherhood were helping me do my job."

Based on what they learned during the participatory listening process, Napawan and Snyder designed a suite of communication materials, including flash cards that provide reminders of the kinds of kitchen waste that contribute to the overall waste problem, marked manhole covers on city streets, and related graphics on wastewater trucks. Together, the trio of bright-green and integrated graphics served as a reminder of the connection between individual kitchens, city infrastructure, and the larger ecosystem of the San Francisco Bay Area.

In the design process, Napawan and Snyder developed valuable relationships with community members as they worked together over a three-year period. They hope their work can provide an example for other municipalities to consider small-scale design interventions and community-based participatory approaches to urban environmental challenges.

Napawan acknowledges that this work is not without risk. She says, "Once you start listening, you don't know what will happen, you don't know where the conversation will go."

Professional Listening

Across fields and professions, listening tends to be a more implicit rather than explicit focus and rarely addresses power directly. Some are working to change that. This section provides a series of interviews, narratives, and vignettes showing listening in practice. Like all material in this book, they are not intended to account for the views of entire fields but instead are presented as examples from which science communicators might learn.

Listening like an Ethnographer
As Brooks and Olson's method demonstrates, ethnography is important in both research and practice in the social sciences, originally emerging from anthropology. While ethnography can be controversial—for reasons ranging from the method's colonial origins to its lack of reliability to its use of exotifying tropes[12]—contemporary ethnographic approaches are attempting to respond to some of those critiques.

Amber Wutich is an anthropologist at Arizona State University and a methodologist with expertise in ethnography. She says that despite the central role of listening in ethnographic interviewing, it's not necessarily a subject that receives enough attention or instruction.

Reflecting on her own work, she recently wrote, "In addition to working with marginalized communities to better their situation in ways that are important to them, I think there is one other important ethical obligation that researchers have when working in impoverished and exploited communities. That ethical obligation is to actually listen to what people are telling you."[13] She notes that in her research on water insecurity, she asked people why it mattered in their lives. She was surprised that people frequently talked about their feelings rather than water shortages, costs, or sources.

"People told me about the enormous distress, their suffering, and their anguish about the water situation. Now this was totally unexpected—there was no theoretical guidance in the literature that had led me to expect these kinds of conversations, and I didn't feel well prepared to investigate water-related distress." But these conversations led her to go where the listening took her. She now says she is "careful to listen to what people are telling me matters and let that guide my research—even if that takes me in directions that I find unfamiliar or scary."

Listening like a Lawyer

Similarly to ethnographers, Gail Silverstein, the law professor from chapter 4, says that lawyers often first learn to listen when interviewing clients. She says that her main focus with students as they begin to interview is to focus on building rapport with clients before launching into any information-gathering effort. "You need to develop a strong foundational relationship with your clients. While some people might argue that your first job is to get the facts, I think that has to come second to listening and relationship building, particularly when you are working with longer-term clients. The facts will come, and in my experience, you get better information if you can first get to a trusting relationship with clients."

Silverstein says that in order to do that, at least initially, she advises believing clients, full stop. "Especially in that first interview, I want students to be in complete belief of their client—that's what helps build rapport." Simultaneously, she says lawyers have to learn to balance listening and believing what they are being told with times that require them to ask questions that might be more confrontational. "A big theme in our clinical law courses involves empathy, attaching to your client, hearing their perspective, and trying to understand where they're coming from. But then you have to step back a bit and see things from a more detached and objective viewpoint so that you can serve them. You need both; you can't just have one or the other. It's a lot for students to learn and takes a lot of practice with many clients to master."

Silverstein says teaching interviewing is so complex that it's hard to convey in a lecture, so she has students do mock interviews so that they can analyze how they are listening. "Listening is not passive; it's listening for information but also for emotion, which is a kind of meditative listening. You're listening for interests, needs, constraints, limits, priorities, and goals. You're listening for what is underneath the words, not necessarily what's being said. We encourage active listening techniques so that you can show that you're listening, which helps the client feel seen. We also teach the importance of silence and nonverbal communication because so much communication is nonverbal."

Listening like a Doctor
Although there have been efforts to help doctors become better listeners, most practitioners argue that this skill hasn't received enough attention. Juliet McMullin, the professor of cultural and medical anthropology from chapter 4, says that while students do talk about and practice listening in what might be called "doctoring" classes early on, it becomes much harder in practice when there are more tasks to manage.

These constraints are reflected by Rebekah Fenton, the Adolescent Medicine Fellow at Lurie Children's Hospital of Chicago, also from chapter 4. She says that as a University of Pennsylvania medical student, she learned one of her biggest lessons in her first year of training. That year, she spent afternoons after traditional lectures with other students who would convene in small groups to work with what is known as a standardized patient. The "patient" is recruited and trained to act out a set of symptoms that can help students learn experientially, practicing communication and other crucial skills.

One afternoon, students were told to practice simply listening to the patient without intervening. Immediately, Fenton found herself offering advice and was called on it. She says, "I'm a fixer, and it made me uncomfortable to not jump straight to trying to solve the problem the patient was having." Fenton realized that she was uncomfortable when she couldn't give advice but that it actually made the patient much more comfortable when Fenton could simply listen first. She says this was a huge lesson that in many ways has shaped her medical interests. Fenton says she now knows how important it is to "hold space"—to be present without judgment—for a patient to share what is happening for them. "I've come to recognize listening as a special role for doctors, particularly because patients may not have other places to talk about what they can with a doctor."

Fenton says that despite the focus on listening in her first year of medical training, it became much more difficult by year two as her medical knowledge grew and the complexities of doctoring emerged. During the intake process, she says, doctors have to get patients to answer on the order of fifty questions about their medical history and will start to combine questions and sometimes gloss over things that might be important to patients while trying to get through the information.

Once, Fenton was going through the intake process with a woman. The woman noted that she'd had a breast cancer scare but had ultimately been told it wasn't cancer. All Fenton saw was "not cancer" before moving on. The patient later told her that she wanted Fenton to at least acknowledge that even the scare had indeed been scary for the patient.

"Part of what we are also trying to work with is hearing differently and to think about listening as rapport building. Not everything will be understood in a single visit, but you can begin to establish a relationship, to keep the conversation going," says McMullin. However, she empathizes with the reality of medical practice for many doctors. "The idealized narrative medicine process, which is incredible when it can happen, is hard in a standard setting. Yet there are ways that we could and should be incorporating better listening into medical practice."

Listening like a Journalist

Wudan Yan is a freelance journalist based in Seattle, Washington, who has covered topics ranging from the refugee crisis along the border between Myanmar and Bangladesh to the legacy of nuclear testing on Native American communities in the American Southwest. She says there is no doubt that listening is a big component of being a journalist. However, in her experience, the emphasis on it has been more implicit than explicit. "I'm not sure how journalists separate listening from any other part of the work; it's just so fundamental. At least for me, keeping an open mind and truly listening is where the best stories come from."

She says when she thinks about learning to listen well, it came from a meditation practice she developed before becoming a journalist more than from any formal learning in her profession. Yan says, "I didn't come to journalism through traditional journalism training. I didn't study writing as an undergrad; I wasn't at the college paper. I didn't go to journalism school or a science communication program." But, she says, in her early twenties she joined a yoga studio and developed a meditation practice that first taught her to listen to herself deeply, a discipline she says has been crucial in learning to listen to others.

Lizzie Johnson, a journalist who has been covering some of California's most devastating wildfires for the *San Francisco Chronicle*, says she was lucky to take an interviewing course in journalism school at the University of Missouri where listening was a key lesson. "I learned pretty quickly that listening isn't just about asking the right questions. It's about watching for body language and allowing quiet to seep into the conversation even when it feels uncomfortable. Because it's oftentimes in that hollow space that people feel more comfortable or more compelled to talk."

She says that in an interviewing course taught by Jacqui Banasyznski, a Pulitzer Prize–winning journalist, students would be given assignments such as going out to find the first person they could and complete an interview in ten minutes. "We never actually wrote the story from our interviews. But we were learning about asking good questions and whether we gave space for someone to actually talk."

Banasyznski also gave students the chance to interview her, providing them with firsthand experience in how questions could either shut an interviewee down or offer them a chance to provide a more expansive response. "If we asked something that had a yes or no answer, it would just become so obvious that the conversation was over. But then you'd hit on these questions that started with things like 'How do you feel about x?' and you could feel the difference," says Johnson. "There was some emotion in the learning of it, a little bit of shame if you asked the wrong question and shut things down, but you got a feel for getting a real answer based on a real connection when you asked the question in the right way."

Listening like a Therapist
It is, in many ways, the central job of a therapist to listen. And it is one of the few professions that dives in, professionally, to what it means to listen well, to intentionally center listening. However, much as in other fields, listening is an evolving process refined through practice for many therapists.

"The way I approach listening is subtle, and I try to follow my intuition. It's a whole-body and whole-sense experience. I try less to think about it than to open to it," says Leslie Davenport, the practicing psychotherapist and author we met in chapter 4. She describes mirroring and open-ended questions or requests like "say more" as techniques that can help to encourage conversation. "After many years of practice, I listen with my whole self while also mentally tracking words and phrases that might provide some insight. Listening with my whole being open is important for building rapport."

Davenport notes it is also important to understand when listening may not be enough. She says, "It doesn't always serve a client for me to continually listen to them repeat the same story. That's when the rapport building comes in—when you are able to take the trust that has been developed and help them begin to connect the dots between their struggles and the choices they are making. We all have blind spots where we don't recognize the ways we are contributing to our own distress. When the therapeutic alliance is strong, clients are able to be more receptive to feedback that can be emotionally tough to face."

In the end, though, Davenport sees her role as one of helping clients move toward their own decisions, of witnessing, and of providing a sense of safety and wholeness as people work through difficult times. That also means having to work with her own feelings and reactions as they come up in the therapy process. For example, she notes that it can be tricky to provide validation for a client when the subject matter or the interaction runs counter to her own values or mirrors dynamics that are occurring in her own life. "I have to trust the relationship myself, trust what emerges between me and the client."

Both Silverstein and Davenport point out crucial aspects of listening practice, including the value of building rapport and a trusting relationship as the foundation of doing the hard work of questioning what is being heard in an effort to develop answers, together. Even with trusting relationships established, however, listening in a science communication context also raises questions about ethics, accountability, and power.

Listening Responsibly

> I was in the water for quite a while, and by the time they started pulling me back, I was waist deep in the water that had already started coming into the car.
>
> —Cheryl Mickelson, Gays Mills
> Driftless Writing Center, 2019

Stories from the Flood: A Reflection of Resilience is a small booklet filled with gut-punch quotes.[14] Throughout its pages, residents of Kickapoo and Coon Creek watersheds in southwestern Wisconsin share their feelings about the now persistent flooding they are facing with a blend of awe, fear, sadness, and resolve. It's an area used to several big floods a century, but nothing like that of the past fifteen years.

"The Kickapoo River and Coon Creek watersheds are beautiful. They're filled with steep hills, bluffs, and flowing rivers. Flooding is a somewhat common occurrence, but it's been exacerbated by climate change," says Caroline Gottschalk Druschke, an associate professor in the English department at the University of Wisconsin–Madison.

Druschke, recently returned to the Midwest, had been spending time developing relationships in the area and taught a community-based learning class with the Kickapoo Valley Reserve. "Because of that, I was contacted by the Driftless Writing Center, somewhat out of the blue. They had an idea for an ambitious project to collect stories about catastrophic flooding in the area. They were applying for a grant from the Wisconsin Humanities Council and needed me to serve as a 'humanities expert.' I wouldn't call myself that, but I'm committed to using my university power to help where I can, so I told them to put my name down," Druschke says.

"This was a listening project. Their idea was to give people a chance to tell their stories to an engaged listener—somebody who would cry with storytellers, laugh with them, and bear witness to what they went through. This is a huge ask in a community where people are pragmatic and generally suck it up. Upper midwestern farmers are like 'I'm fine. I lost my house and my barn and fifty cows and a tractor. But you should talk to my neighbor who suffered.'"

Druschke is particularly equipped to work in this kind of community setting. She completed her undergraduate degree at the University of Iowa in social work and a doctoral degree in rhetoric at the University of Illinois at Chicago. She says both fields, though seeming very different, focused on interpersonal relationships and dialogue, as well as radical critique of inequitable structures, albeit from different angles.

She is also sensitive to the many complexities of community-engaged work.

> In graduate school, I was involved in a participatory science project that turned out to be a total bomb. We somehow thought people would appreciate science simply by participating, but instead, they thought, 'Scientists just want our data.' I am also wary of storytelling as an extractive endeavor, especially with trauma, which people like Eve Tuck and K. Wayne Yang have written about beautifully.[15] We had to walk a fine line with this project. I had been invited in, and that helped. I was not proposing a project; I was assisting my partners in carrying out work they wanted to do.

Druschke says when they learned they had received the grant, along with another, "we decided to get students involved in the project. They were great listeners; they were deeply engaged. We were a part of the best community-based learning experience in history! The students were sobbing at the end of the semester; some are still contributing to the project. A lot of that was organic, and there was a ton of goodwill. The students helped prepare the booklet that was distributed at an event with community members and political representatives—everything was magnificent."

In some ways, the success of the project is where things got harder. Druschke says there was a lot of interest on campus after the university released a news item about the project's success. More people wanted to get involved.

> It was a classic "good intentions gone awry" thing. A class with more research-focused goals joined in, and everything necessarily became more formalized. Oddly, with the COVID-19 shutdown, things slowed down in a good way and let us hit the pause button.
>
> What I've realized is that I have long-term commitments in that area, but others on campus weren't necessarily aware of that.

I've had to reckon with the idea that the setup of the university is in many ways not compatible with long-term community engagement. With classes, we work on a semester basis, and there's no long-term accountability. So, I'm the person that has to remain accountable. I have taken on the responsibility to build a multiyear relationship with this group, which I expect to continue indefinitely into the future. That's what allows me to ethically bring students in and out.

Ethical and Accountable Listening

As the narratives throughout this chapter indicate, when science communicators take up listening as a crucial practice, a new set of ethical questions emerges. One initial step in the listening process might be cultivating a willingness to be changed. However, when steeped in relational theory, listening must be about much more than how it affects an individual listener.

Listening well raises the question of how responsible the listener is for what is heard. What does the listener do with what they learn? On the one hand, listening without action can be an empty exercise. On the other, the listener can risk being extractive or appropriating knowledge. In other words, there are the ethical dimensions to listening that must be addressed. Some potential ways forward come from others who are thinking about how to make the process less extractive.

Johnson, the journalist, has a series of steps that she takes to be as accountable as she can during her reporting. "As a journalist, I never want to go into an interview that leaves the person worse off than when I got there. That is fundamental for me." She adds, "Over time, you learn your own system for how to approach people in a way that is healing or as cathartic for them as possible. In part, I do this by listening well but also by laying out the groundwork for the reporting up front by saying things like 'I'm going to record this interview, I'm going to be writing notes in my notebook, if you ever want to see the notes, please let me know. I have nothing to hide, and I'll be accuracy checking everything with you at the end of this.'"

Julian Reyes, from chapter 2, says that while there's more recognition in the science community about the need to listen, it's not as fully realized as it could be. "I think the advice to listen is definitely reaching some scientists. But there's also a question of where and how we're listening. A lot of us probably need to be thinking more about meeting at other people's tables and at their invitation. You have to do the work to build trust with people, not just swoop in and quickly leave. I try to think about it as a long-term relationship, which is hard given funding rarely covers that kind of thing, but I still think that mindset is important."

Yanna Lambrinidou, the anthropologist in chapter 4, has both written about and taught the value of ethnographic listening in science and engineering.[16] She views listening as, first and foremost, a tool for learning technically and morally relevant information about complex problems people experience in their day-to-day lives. For her, listening is a necessary skill for competent professional practice. After years spent on the frontlines of the lead-contaminated drinking water catastrophe, Lambrinidou has seen the downsides of scientific expertise in crisis situations firsthand. She feels strongly that the value of listening is sometimes "diluted to become about empathy and compassion, when we're talking about justice and accountability." Not only that, but who is heard and how are crucial concerns.

Lambrinidou says that "how scientists and engineers listen is inseparable from questions of power, because it can be done in ways that bolster community knowledge and strengthen scientific understanding of real-world problems. But it can also be done in ways that overlook or distort community knowledge and compromise scientists' ability to help people in ways that are scientifically sound and experienced as desirable and just."

Her experience led her to co-teach a class on ethics for scientists and engineers at Virginia Tech. In the course, Lambrinidou led with the premise that as scientists or engineers, students cannot possibly possess all information required to help address complex problems affecting diverse communities in multiple and interconnected ways. Lambrinidou says, "The course was set up that way to enforce, from the get-go, that expertise is relative, that the technical knowledge taught in the classroom is not always complete, correct, or adequate, nor does it comprise the entirety of what scientists and engineers *need* to know—technically and morally—to act responsibly."

She had students go through a three-step exercise that involved developing the ability to hear other people as expert sources of information. Lambrinidou stresses that without this kind of ethically based listening process, practitioners are vulnerable to negligence and the potential to actively cause harm because "they can be quite insular in deciding what 'good' is, which leads to the question of accountability and who ultimately decides that the 'good' was indeed done."

Listening can also be an extractive process, and Lambrinidou says that it is particularly thorny because exploitation and appropriation can occur on many levels. "If practitioners can embrace the idea of truly offering work in service of communities, can manage to untie their contributions from the academic achievement model that glorifies individualism and heroism, particularly as defined by scientists and devoid of community input, there is hope. In a truly mutual relationship, there can't be a hero."

"We have to begin to enter consenting agreements with the foundation that all parties have knowledge to offer and all rewards are distributed or otherwise negotiated," says Lambrinidou. "Communities ought to be able to safely terminate collaborations that don't meet their needs, and researchers ought to know when to remove themselves from projects that frustrate them to the point where they start treating people with novel questions, experiences, and knowledges as somehow 'defective' or a 'threat.'"

Lambrinidou says one of the major promises of ethnographic listening is using it to explore contested knowledge, not only to describe a person, a culture, or a belief system. Listening to people on the margins can help experts begin to understand the basis of knowledge claims that, at first sight, might seem improbable or even preposterous. From there, it can support practitioners in reevaluating potentially unexamined assumptions, holes, or errors in their own knowledge and can guide them to make important adjustments to their practice. "There is a broad conversation to be had on the ethics of engagement by STEM professionals," she says.

As the many viewpoints in this chapter indicate, listening is far from an easy, benign, feel-good process. In fact, listening well often results in more questions than answers. And on the many high-stakes, emotional, and contentious topics that scientists and communicators navigate today, conflict is inevitable.

Takeaways
- Listening has been an overlooked focus in science communication work, particularly when it comes to power and accountability.
- There are many ways to approach listening, and different techniques that range from active to deep listening make sense in different contexts.
- Many science communicators are already experimenting with listening techniques and have many lessons to offer.
- An underappreciated question about listening in science communication is what practitioners do with what they hear.
- The ethics and accountability of listening processes need better scaffolding in science communication to help ensure they are not extractive or harmful.

Questions for Exploration
- What issues did reading this chapter raise for you?
- How are you currently listening in your work?
- What are some new listening practices you could incorporate?
- Whom are you listening to and how?
- What are you doing to address power in the listening process and ensure you are accountable to the people you are listening to?

CHAPTER 6

Working with Conflict

CONFLICT IS INHERENT IN RELATIONSHIPS. Although challenging, it is also an important ingredient in growth and change. Unfortunately, discomfort with conflict can lead to calls for "civility" as conflicts come up. Yet civility can be code for compliance in many contexts, and attempting to suppress a conflict too soon can lead to a false sense of agreement, ensuring that conflict is certain to come up again.

Accepting, even valuing, conflict as a necessary part of large-scale transformation has the potential to change scientific engagement on controversial subjects. As science communicators begin to engage more deeply with conflict, they may come to see it not as a flame to be fanned or suffocated but as a resource for better understanding one another and the issues at stake.

Conflict is an obvious presence and ongoing process in science communication, though one that has generally been viewed as something to be ignored—or in some cases, emphasized—rather than simply an important part of moving forward. Even the sense of what a conflict is can be highly subjective—one person's conflict can be another's way of doing business. Some people gravitate toward being in "the fight," and others tend to shy away from it—either way, it's two sides of the same coin, at times giving conflict a much larger role than it needs to have, whether through neglect or inflammation.

The experiences that follow demonstrate how valuable it is for science communicators to enhance their ability to navigate conflict. Not addressing its role or equipping communicators to work with it has been a major oversight in science communication efforts; all the evangelizing in the world for the value of engagement is not going to help without directly addressing the discomfort—or on the other side, potential harm—that can come from being ill equipped to work with conflict. However, personal comfort with conflict is only one piece of a much larger picture. In some cases, it comes to entail power struggles that scientists and communicators—particularly those who are precariously employed or otherwise vulnerable—can find deeply challenging to address.

Waiting for Rain during a Drought

Even something as common as waiting for uncertain news can lead to conflict between people who tend to view the glass as half empty versus those who view it as half full. What happens when that uncertain news is about something completely out of our control, such as the weather?

Dan Macon is both a shepherd to his own flock and a livestock advisor with the University of California Cooperative Extension program.[1] Macon's livelihood is tied to the land and to water, which is a vital ingredient in creating the unique grasslands his animals depend on. As a small-scale sheep rancher in the foothills of the Sierra Nevada, he spent a lot of time waiting for rain during the state's most recent long drought.

Kate Sweeny, professor of psychology at the University of California, Riverside, studies the kind of waiting that Macon is faced with—that is, waiting for uncertain news. Sweeny and her colleagues found that while waiting for uncertain news, people may focus on preparing—emotionally and logistically—for any possible outcome[2] and that there are things people can do to "wait better."[3]

People also tend to shift between optimism and pessimism, and while both states can help increase readiness, they can also be in conflict. Optimism encourages people to take preparative, proactive actions, and pessimism helps people to prepare by protecting themselves psychologically from worst-case scenarios. People with different waiting styles may therefore end up disagreeing about how to respond.

Although Sweeny's work has focused on individual experiences of waiting for things with definitive outcomes, like difficult medical diagnoses or news of a job, she thinks her research findings can extend to waiting for a resolution to any stressful uncertainty, like waiting for rain during a drought. The differences between waiting for something that directly impacts an individual and something like drought that, for many, has more indirect impacts are likely to be found in the ease with which people can distract themselves and manage their expectations.

"It is easier to distract with something distant and abstract than with something imminent and personal," Sweeny says. In other words, the less personalized the experience, the more optimistic people are likely to be about the outcome. In the case of drought, therefore, it may be harder for farmers, ranchers, and others whose livelihoods are personally impacted to remain hopeful than it is for people who know water will most likely come when they turn on their tap.

Josh Reynolds, a veteran and licensed mental health counselor who works with military members and their families, is also well versed in waiting. Waiting for news about deployments or extended periods of combat is simply a part of military life, but that doesn't make it easier. "Waiting can be deeply unwelcome, and many of us resist it. We do things to regain a sense of control and avoid the discomfort of waiting, which can get in the way of experiencing and expressing anxiety and grief, and that may be counterproductive," says Reynolds.

One of the more complex aspects of understanding the role waiting plays in something like drought is how collective emotions—common feelings that surface for groups of people as a result of shared experience—come into play and how they can lead to conflict. However, waiting together can also be comforting. As Sweeny says, there may be more space for collective conversations with something like drought that affects a broad group of people rather than an individual.

In cases like this, understanding that people can experience the same event differently and have different coping mechanisms might help ease conflict. Macon, who has contributed to many community workshops and started a virtual group for ranchers impacted by the drought, says, "Waiting for it to rain is easier when you know other people are waiting along with you." However, even something as simple as different perspectives about uncertainty can lead to conflict.

Understanding Conflict

Conflict is an inherent part of interactions among individuals and groups that are working together, and it is bound to arise during the engagement process. If handled skillfully, conflict can lead to fresh insights.

Personal, sometimes even physical, effects are evoked when we brush up against the "other." Yet by accepting conflict as a natural part of an engaged approach, science communicators can develop practices that help them navigate it more skillfully.

There are many different theories of conflict. One school of thought was summarized in a special issue of *Science*. Researchers describe conflict as stemming from the "otherization" of individuals, groups, and even the natural world.[4] This otherization leads us to more easily empathize with our "ingroup" and to dehumanize "outgroups." While identification of the other is important from a development angle because it helps us to form our identities as children, it can become deleterious as we mature because it leads to an emphasis on differences between "us" and "them."[5] There is also a growing body of work on environmental issues like climate change as precursors and additive stressors in larger societal conflicts.[6]

At an institutional level, there are well-researched traditional conflict resolution methods including adjudication, arbitration, and mediation, as well as newer ones like restorative justice. There are also many new and interesting techniques for working with intractable conflict, such as situational reframing, which focuses on changing the dynamics of a situation rather than the outcomes and reintroducing nuance, complexity, and compassion into the process of understanding each other.

At the individual level, there are many resources related to enhancing the ability to withstand and effectively participate in situations where conflict is present. There is the rich literature around the concept of nonviolent communication, which we explored in chapter 5. In addition, psychoanalytic literature provides a deep set of resources on the role of conflict in interpersonal relationships.

Contemplative practices from a wide variety of traditions are likewise useful at the individual level. Many contemplative practices focus on helping practitioners remain calm during stressful events. Indeed, empathy- and compassion-focused practices and exercises are helpful, though it should be noted that they also have their limits, particularly when they focus solely on individual responsibility to deal with what can be structural matters.

Learning to work with conflict is a difficult proposition. While conflict manifests in interactions with others, much of the real work of tolerating conflict—indeed, of being able to sustain a career working on contentious topics—invites introspective work with internal conflicts. A real paradox of conflict is that while it can be highly uncomfortable, it also provides opportunity. Accepting and even valuing conflict as a natural outgrowth of an effective, engaged approach changes the skills science communicators may want to cultivate.

Facilitating Dialogue on Wild Horses in Rural California

Laura Snell is a hard person to forget.[7] Her dog, Zuri, is a constant companion, one she deemed better than a firearm for field protection, which immediately tells you something about her work.

Snell was raised in Iowa and trained as a rangeland scientist at the University of Nebraska before finding herself working in rural northeastern California. Modoc County is bordered by Nevada to the east, not far from the Black Rock Desert of Burning Man fame, and Oregon to the north, a couple hundred miles from the Malheur National Wildlife Refuge. Like so many border areas, it has its own unique culture and community. It is almost the complete opposite of most people's mythic sense of California beaches, palm trees, and liberal values.

It wasn't long after Snell started as a rangeland cooperative extension advisor with the University of California Cooperative Extension program that she found herself in the middle of a vast controversy. In recent years, Modoc County has had a boom in its wild horse population. The population is in direct conflict with the many ranches that graze cattle on the same public lands that the horses roam.

Because watering holes can be hard to come by in this high desert country, they are a material site of conflict. Pronghorn, deer, cattle, and wild horses are all visitors to the springs and ponds scattered across typically dry grasslands. The number of wild horses has jumped quickly in recent years, bringing a host of water-related challenges and no small amount of controversy.

Wild horses, today descended from domestic horses, are covered under a federal law focused on coexistence with livestock and wildlife. That has meant rounding up wild horses with some regularity because they have few natural predators and their populations grow quickly. However, legal and financial woes have stalled wild horse management over the years, with horse "gathers" taking place infrequently.

As a result, in one part of the Modoc National Forest known as Devil's Garden, there is a fluctuating population of wild horses that numbers in the thousands in an area estimated to sustain only hundreds. While cattle tend to be closely monitored and rotated to different pastures, that is not true of wild horses. Therefore, the rise in population is hard for both livestock and ecosystem managers; horses crossing local highways are even becoming a problem for transportation officials. Contraception programs are complex, Snell notes, because the treatment requires closer proximity than the horses are willing to allow.

To better understand what has been happening on the land, Snell and her collaborators with the US Forest Service hike Modoc and Lassen Counties installing wildlife monitoring cameras. Snell also measures vegetation and stream bank changes several times a year. But monitoring and collecting and analyzing data are only one side of the job. The other is navigating the human emotions attached to wild horses. From ranchers to wild horse advocates, and the whole spectrum between, people tend to have strong feelings about the horses.

"We sometimes have ranchers who are not allowed to use their permitted allotments at all—and many more that have proportional cuts—due to the large number of wild horses. These cuts affect our local economy and relationships with the government. And the cuts to cattle allotments still don't provide respite for degraded ecosystems," Snell says.

Engaging with wild horse issues has meant learning new skills to complement her scientific and technical expertise. Snell recently took a facilitation course to better prepare to guide meetings and help groups with differing opinions work together. She says, "I'm trying my best to listen first and then to focus on the best science available."

A couple of years ago, Snell partnered with colleagues at the Modoc National Forest to foster dialogue between diverse groups concerned about wild horses. "Bringing people together who are on completely different sides of things and seeing them talk and find common ground can be rewarding," says Snell. On a field tour, participants walked together to what many agreed was a degraded spring site. They found about sixty wild horses that had started to occupy the area.

As the group sat at the spring and talked about the ecosystem and economic impacts of wild horses, emotions began to run high under the hot afternoon sun. It was clear that this would be the first conversation of the many that would be needed, but the meeting did allow for some new insights and relationship building. Snell says she can sometimes leave these events a "bit frustrated and exhausted but also encouraged to reduce misunderstandings around the research on these horses."

Since that time, the wild horse situation has only gotten more heated, and Snell sometimes worries for her physical safety and what it will be like to raise her family in a small town, working on something that makes her a target. "I think it's difficult to understand the complexity and scale of the wild horse issues in the western United States. It is not just about horses; it's about rural economies, wildlife habitat, and multiuse public lands."

Snell says there are times her job keeps her up at night, but she keeps going despite the risk and potential for burnout because she has a deep dedication to the place she calls home. She wants to continue working with ranchers and have her children enjoy public lands, including the horses that are on it.

Like Snell, scientists and communicators are increasingly finding themselves in the middle of controversies they were not prepared to work with. In Snell's case, she has worked to reduce otherization in her community with dialogue and facilitation, helping the group members to better empathize with one another. However, she also sees herself facing the challenges of a conflict with no end in sight.

Scientific Issues as Intractable Conflict

Peter Coleman is a professor at Columbia University and the director of the Advanced Consortium on Cooperation, Conflict, and Complexity.

Coleman works on intractable conflict, which he defines as long term, chronic, polarized, and not easily addressed using mediation and other common conflict resolution techniques. Some of the best-known examples of intractable conflict include Israel and Palestine and the Troubles in Northern Ireland, but Coleman argues these kinds of conflicts happen all the time in situations that involve everything from local communities to family systems.[8]

Many scientific matters are also taking on the nature of intractable conflicts—long and drawn out, highly polarized, largely immune to tactics like negotiation. Climate change, fisheries management, and water resource allocations in arid places are just a few examples.

However, the label of intractability is less valuable than exploring some of these topics through the lens of conflict, simply because it changes the kinds of solutions that might be considered useful from a communications standpoint. Coleman's work has led him to focus on "dynamical conflict resolution." During intractable conflicts things become incredibly polarized—there is only black and white, wrong and right, us and them—and a potential solution is to reintroduce nuance and complexity into the situation to begin to "break open" the two opposing perspectives.[9]

Introducing, or even simply allowing, nuance and complexity or discussions of the meaning of information into a science communication or engagement effort is a bit counterintuitive. Most training in science communication is about simplification and streamlining, distilling complexity down to a couple key points to be repeated, focusing on more accurate information as a solution. That approach may work well in some cases. But what if it heightens conflict and mistrust in other situations? It seems worthwhile to take a step back and think about whether complexity, either in information or emotion, might at times actually be helpful in science engagement, particularly where ongoing conflict is present.

Complexifying Conflict

Fifteen black-and-white photos hang on a wall.[10] One shows the tiny figure of a man inside a large concrete pipe slated to become part of California's vast water infrastructure, another a man rendered so small in comparison to a finished canal filled with flowing water as to be barely perceptible. Next to that display is another black-and-white photo of an enormous dam project, taken in the 1960s. The promise of that photo stands in contrast to the contemporary color photo of the same dam with the kind of dry ring around the top that has become a ubiquitous symbol of water shortages.

"California's water culture is grounded in the weather and our wet and dry cycles along with the fact that you couldn't have built our major cities without water. It's a culture that is dependent on water," says Rina Faletti, the art historian, environmental humanities scholar, and exhibition curator who curated this art exhibit on industrial photography and the state's massive water infrastructure. "One of the overwhelming things that looking at these industrial photographs does is reveal how absolutely enormous California's water supply infrastructure is. The monumentality—a term I usually shy away from but that is appropriate here—of the engineering work, the money, and the political will that went into creating it is almost unbelievable."

Despite the vast investments in water infrastructure, in a place with more arable land than water, the reality remains that for some, enough water will never be enough.[11] "Water is a cultural commodity. And I don't mean it in the sense that I can buy a bottle of water. I mean that it informs California's identity. We're always teetering on that edge between scarcity and excess. What's enough? What's not enough? We develop water supplies that create the need that the supply is purporting to meet. It's a conundrum."

Despite the amount of research that goes into answering questions about water use and conservation in the state, Faletti says that it's a cultural question and one that projects like the one she has undertaken can answer outright, or at least help to get people talking about water in new ways. She says using "boundary objects" like these photographs can open up new conversations about long-standing conflict. "For instance, when comparing industrial and recreational water uses, a nuanced discussion reveals that recreation takes place on reservoirs or rivers that are highly controlled and managed by a massive system of utilitarian water use infrastructure. Furthermore, that engineered infrastructure is sometimes designed to look like neoclassical architecture. In that context, I'd want to explore that visual reality and where utilitarian use meets cultural use.

"I'm curious about how we pose and discuss these kinds of questions in a common language so that everyone—including consumers, managers, builders, engineers, politicians, historians, schoolchildren, farmers, and farmworkers—can sit in the same room and understand the complexities together," says Faletti. "We need language and images that allow us as a culture and society to get the scale of the problem—how big it is, how it affects us—so that we can all envision our roles in possible solutions."

Faletti's approach of using art, in this case industrial photographs, to create a common language amongst a group of people who might normally struggle to talk is just one example of a way to introduce more nuance into a highly polarized debate. There are times, however, when better conversations are tested as a method for working with conflict.

Power and Conflict

Linda Estelí Méndez Barrientos, a doctoral candidate at the University of California, Davis, puts it plainly when she says, "I think focusing on better communication is distracting. There are these much larger issues of power at stake, and it's convenient to have a lot of very smart people worrying instead about communication." She refers to climate change as an example of a topic where decades have been spent enhancing communication efforts by the science community that could perhaps have been better spent addressing its root causes.

"Power is complex. It can be economic, political, or discursive. The science community tends to have a high level of discursive power but usually less economic and political power, relatively speaking. However, not every science communicator has the same level of discursive power—you can see who gets to speak for science and who doesn't. And, with something like climate change, you can see how some prominent science communicators have in many ways set the discourse. Yet discursive power is not enough to overcome power imbalances in other arenas. It's not a model of social change on its own."

In her own work, Méndez Barrientos is working to better understand how power is negotiated in the environmental arena. She says, "I try to understand how power differences shape policy processes. In the past, we assumed that whatever was written into law or policy was directly translated into practice. However, the gap between policy design and implementation is huge. I'm particularly interested in how inequity and inequality help explain that gap."

She is currently examining power in the context of California's groundwater management. During the state's long drought in the mid-2010s, water availability was a source of much anxiety and conflict. California is a place with a distinct dry and wet season that is entirely unlike much of the rest of the country, so concern about water availability is a background condition for life in the state. When surface water supplies for irrigated agriculture diminished as the drought dragged on, farmers made up the difference by pumping groundwater. Concerns over how much water was being withdrawn, and the related sinking of the ground in response, spread locally and across the state.

In response, California set up a long-overdue statewide groundwater management effort. Groundwater sustainability became a big worry throughout the state during the drought, but some places were experiencing more rapid declines than others, so a phased timeline for implementing the regulations was put into place. That set in motion a decades-long process involving individuals and institutions at many different levels and, by design, leaving many decisions to be made at the local and regional levels. The flexibility has meant that groundwater management is not uniform across the state but rather a highly complex, localized process with different outcomes.

"Groundwater management in California is one of the world's largest-scale policy experiments on collective action to manage natural resources. Disparate access to water resources in some areas makes it the perfect case study to better understand power differences," Méndez Barrientos says. She spent hundreds of hours observing how groups set up to manage groundwater in areas throughout the state negotiate power and conflict, noting that in reality, groundwater users face many barriers to participating in the management process, but these barriers are not the same for everyone. She says that in a statewide survey she and her colleagues administered, they found that "tribal groups and disadvantaged communities, in particular, perceived challenges to participate in the policy process."[12]

The survey also revealed that people generally located information about the process with relative ease. However, Méndez Barrientos says, "this contrasted with perceived lack of opportunities to comment and express opinions and even less opportunities to be adequately represented and influence decisions." This finding, she notes, suggests that access to information is not as big a barrier as access to deliberation among groups involved in the process:

> The power dynamics were set up from the beginning—the way that new administrative units were formed on top of existing power structures means that you're going to keep favoring

> the people that already have connections and money. The people who the management process was ostensibly set up to protect—people who depend on groundwater for drinking, bathing, cleaning—are not being included. In this case, it is the large agricultural and urban water districts, and they each have their own set of discursive, economic, and political powers. It's very hard to address these inequities that are baked into the system. We're seeing that power is working to protecting the status quo and not really meeting the policy agenda that was set out.

This conclusion reflects sentiments shared by Mario Sifuentez, an associate professor of history and director of the Center for the Humanities at the University of California, Merced. "I grew up in a migrant farmworker community in the high desert of eastern Oregon.[13] When I first came to California, in some ways it felt familiar and in other ways it didn't. I remember driving down the highway and seeing the 'No Water, No Jobs' signs," he says.

> A lot of them were in Spanish, which struck me. I had a hard time wrapping my head around the idea that farmworker welfare was tied to farmer access to water. Growing up, I'd never encountered the idea that farmers just wanted water to provide a better life for us farmworkers.
>
> I found that the organization behind most of these billboards was made up of groups far from the affected communities. So I started to look at more connections between farmers and water. Of course, conservative media espouse the fish versus people mindset, but during the drought, a lot of liberal media also put out these long articles about how it was affecting farmworkers. They'd write about how hard life had gotten and how people

were waiting in line for food, which was weird because it's not as if farmworkers lived a sustainable life before the drought.

"At the same time," Sifuentez says, "some farmers have seen their profits double, triple, quadruple, and the quality of life for farmworkers is not improving at all. So I've started to say, 'Water wealth does not trickle down.' It stays at the top and doesn't do anything for farmworkers in California."

Indeed, Méndez Barrientos cautions that not having a robust understanding of power can lead to an overemphasis on some solutions over others. "If you look at groundwater, you see that all forms of power are important, so focusing on one will not explain what you are seeing in reality. So, while communication is important, it's part of a discursive battle, and without political and economic power, we're not going to see social and environmental change," she says. "I think there's definitely some truth to the idea of intentionally letting people fight over the discourse instead of the essence of important policies and processes."

Transforming Conflict

More than ever, scientists and communicators are involved in engagement efforts that can put them in the middle of unanticipated conflicts. Although these endeavors to engage are rewarding, they can also be difficult when encountering sometimes vastly different viewpoints, or even navigating the nuanced views between those who generally agree. They are even more complicated by power and how it relates to conflict.

On the surface, many conflicts that appear to be about science are about broader matters. For example, while something like drought may seem to be straightforward, it is not. At a basic level, it's hard to know when one starts and ends, particularly when there are different kinds of droughts ranging from agricultural to the more commonly thought of meteorological drought. Macon's experience of sitting with different ways of looking at a fraught psychological and rhetorical environment presents one layer of potential conflict.

In Snell's case, information about how wild horses affect water systems in an arid area has turned out to play a minor role in a much larger conflict over rural life, one that is likely to stay with her for the remainder of her career. Faletti takes on a completely different way of working with conflict, adding nuance and complexity, creating a way for groups that don't normally speak the same language to manage tensions together. Méndez Barrientos is addressing the issue at the root of many conflicts: power.

Each case demonstrates some of the different ways that conflict manifests and can be addressed, with people who play different roles and see different things. There is no one-size-fits-all when it comes to working with conflict. In some cases, becoming more individually comfortable with conflict or learning to back off a tendency to create conflict is useful. In other situations, understanding and even confronting power and its uneven application and effect are a necessity.

It is important to note that while conflict may be understood as an outgrowth of otherization, dehumanization has also been recognized as a tool, particularly when it comes to violent conflict. In her book *Down Girl*, philosopher Kate Manne has argued that in cases of violence, the very humanity of the person being violated is exactly what is being exploited.[14] In those cases, pleas to recognize the humanity of others in an effort to stem conflict or violence are moot. In addition, author and activist Sarah Schulman has argued that conflict itself is contested territory. In her book *Conflict Is Not Abuse*, she lays out many cases demonstrating that people tend to call things conflicts that are not while allowing true abuses to continue.[15]

Even determining who is conflicted is complicated. Restorative and transformative justice frameworks attempt to heal harms by helping the parties involved understand the ways that we are all involved in conflict—that it's less about one party being harmed by another and more about the many ways that structural problems affect all of us. These are all nuanced points, but anybody who wants to address conflict in their science communication efforts might consider exploring them further.

As laid out in the first part of this book, the science communication landscape, the people doing the work, and the topics they are working on are evolving rapidly. It is a trifecta that presents enormous challenges for science communicators, with implications for communities, particularly on emotional and contentious issues. In addition, conflict is not experienced in the same way by everyone, nor are its consequences universal.

Navigating conflict can feel like walking a knife's edge. One of the best tools that a professional, or even simply a human being, has is to discern when to engage and when not to. For that reason, it is important for individuals to know that it's perfectly okay to rely on colleagues or to walk away from conflict and return to it another day—or not. Those who can return should be open to the prospect that resistance may be strongest precisely when change is most possible.

Takeaways
- Conflict has an ongoing presence in science communication, and while it is recognized, truly engaging with it in a nondefensive way is rare.
- Conflicts can be interpersonal and, using skills such as self-reflection and empathy, navigable.
- Conflicts also have systemic causes, and it is valuable to distinguish between interpersonal and systemic matters in addressing conflict.
- Differences in power or status may be driving a conflict. Sometimes what might appear to be about communication is instead about power.

- Engaging in conflict invites deep discernment about your role as an individual, where you have power or can build it with others, and where you should and shouldn't intervene.

Questions for Exploration
- What issues did reading this chapter raise for you?
- Where does conflict show up in your work?
- In what situations can you work with conflict?
- What kind of support would you need to work through conflict?
- How are you addressing power in your science communication work?

CHAPTER 7

Understanding Trauma

UNDERSTANDING TRAUMA AND HEALING IS CRITICAL for science communicators. Trauma is generally defined as the emotional response to an overwhelming and often terrifying experience.[1] Many people have directly experienced trauma in some form or another. From growing up in abusive or addiction-ridden homes to suffering from serious illness, having been in or witnessed accidents or violent crimes, losing jobs, losing loved ones, fighting in wars, experiencing racism[2] and deep structural inequalities over generations.[3] In these and the many other ways trauma plays out in human lives, most people get it on some level. This is not to say that all traumas are the same, only that trauma is experienced broadly.

There are the wounds trauma can leave, both visible and not. Some have actual scars and others the aftereffects of anxiety, depression, dissociation, grief, insomnia, violence, flashbacks, or bodily pain.[4] And under the best circumstances, healing is also possible—there are resources to be found, sometimes where they are least expected, and people pull together in community with generosity, compassion, and love.

Trauma shows up viscerally in disaster situations. It is particularly easy to recall distinct intense events like Hurricane Katrina[5] in the United

States or the 2004 tsunami in Thailand. In cases like these, the number of people killed or left homeless and mourning is almost unimaginable, and trauma is a large and obvious concern.[6] Recent years have seen devastating hurricanes and wildfires around the world, not to mention the coronavirus, affecting communities as well as scientists and communicators themselves.

When Home Doesn't Feel Safe
"That fire erased our sense of safety, not just at the individual level but across our whole county—it was a community-scale trauma," says Tessa Hill. "Early on the morning of October 9, 2017, we awoke to multiple friends contacting us asking if we were evacuating. We launched ourselves out of bed, turned on the TV and pulled up social media feeds, and began a frantic hour of messaging friends and family trying to understand the wildfire chaos around us,"[7] wrote Hill—a professor at the University of California, Davis, and Bodega Marine Lab—of her experience of the Tubbs Fire.

Deep into that October night, that fire, along with several others, had broken out across her region of Northern California. People in the immediate fire area were jangled awake in the dark, some by authorities and some by neighbors and some not at all, as alert systems across the region failed. A couple survived the night back to back in a swimming pool;[8] the regional hospital was evacuated, and images of elderly and other vulnerable people in stretchers being wheeled to ambulances through thick smoke circulated widely.

The next morning, Hill wrote, "Stepping outside, the air was not just choked with smoke, but large pieces of ash were falling from the sky. By daylight we noticed that some of the ash was as big as our hands, and had legible writing on it; other large pieces of ash falling into our yard were burned fabric. The remnants of our community were falling from the sky."[9]

In the end, the wildfires burned for weeks before being put out, forty-four people died, and thousands of homes burned. Many of the houses that burned in the Tubbs Fire were in the same path as a 1964 fire that had scorched the same general area, only in that fifty-plus-year time span, it had been occupied by much more dense housing, which only served as fuel for the fire.[10]

Although Hill did not end up having to evacuate that time, she complied with evacuation orders for the Kincade Fire in late October 2019; the fire was burning just north of where the Tubbs Fire had. Based on the horrifying experience of those 2017 fires, officials had instituted a series of preventive measures, including shutting off power supplies in the area to avoid sparking electricity infrastructure when high winds and hot weather coincided and made the area vulnerable to another fire. The Kincade Fire started anyway, and in an unprecedented move, officials laid out what they hoped would be an orderly evacuation plan, similar to those for hurricanes.

The scope of the evacuation area sent a shock wave through the state—fire officials were concerned that the fire could jump a major freeway and make a run all the way to the Pacific Coast, some thirty to forty miles away. After watching the fire grow for days, preparing her house and children for evacuation and listening to the police scanners late into the night, the call came for Hill's zone to evacuate.

Under more falling ash and smoke-filled night skies, she and her husband and two young children left their home at 3:00 a.m., driving past the shadowy figures of neighbors filling their cars in the dark. Hill tried to comfort her kids by repeating the mantra "We are safe, we are together." They ended up at a hotel sixty miles away, the first one far enough away from the fire that would also take their many pets. In what Hill describes as a somewhat surreal scene, they were surrounded by others from Sonoma County and all their animals too.

A scientist who studies climate change in ocean environments, Hill is well versed in slower-moving disasters like ocean acidification. However, having firsthand fire experience gave her an entirely new, and more personal, sense of what the future will bring. "I feel like this is a reality that we have to learn to live with, and we honestly don't even know the full shape of it yet," she says. "This year brought home the feeling that it's not just chance; the 2017 fires weren't just a 'one-off.'"

At a gathering of local friends held after everyone who had evacuated returned home, the biggest discussion was about whether to stay or to go—permanently. The evacuation had saved lives but created new concerns. "The conversation tends to run the gamut of emotions. Some people are committed to staying, others are sure they can't keep living in this kind of fear. I swing back and forth—we've committed so much to living here that it's hard to imagine leaving." Hill notes that friends discuss purchasing distant property—though where remains a big question—or vans or RVs that would give them a place to shelter when, not if, these power shutoffs, evacuations, and fires continue to happen.

Tellingly, Hill says she then encountered a small note near the end of a local newspaper article saying that the county school system was considering a move to a year-round schedule, with October taken off instead of summer vacation because it has been the worst month for fire there recently. But who knows how that might shift in the future with what many are already saying is a year-round fire season in the state and now with the coronavirus pandemic on top of it. Hill says, "I feel pretty prepared for disaster at this point, for better or worse. But I think we might just be looking at spending fire season every year being a little afraid."

As Hill's experience indicates, she and others in her community have gone from thinking of large-scale, devastating fires as once-in-a-lifetime calamities to annual events, and their lives are being reshaped by the trauma in ways that remain to be seen. Indeed, as these acute events bump up against one another, many disasters that might once have been thought of as short term are deepening. Recognizing the potential of this kind of ongoing, cumulative trauma is critical in science communication practice related to disasters.

Trauma-Informed Disaster Response
"Just understanding that there is a psychological impact from disasters is valuable,"[11] says Maryam Kia-Keating, an associate professor of clinical psychology at the University of California, Santa Barbara, and a licensed clinical psychologist. Her work focuses on coping and resilience in the context of adverse childhood experiences, trauma, and stress, particularly for vulnerable and understudied populations. She has also researched trauma in the wake of disasters, including a large wildfire that happened in Santa Barbara in 2009.

Kia-Keating has advice for scientists and science communicators who are working with people who have been in traumatic situations, including themselves: "It can be easy to overlook psychological effects of disasters, especially in the early stages, when basic needs obviously take precedence." However, she warns, "neglecting psychosocial well-being is shortsighted. There is a lot of empirical evidence demonstrating its impact on other elements of individual and community health and resilience, both initially and over the long term."

The psychologist says it's critical to understand that recovery does not look the same for everyone, even when they've experienced the same disaster. She notes recovery takes time and that it has many phases, some of which can repeat and overlap. In her research, Kia-Keating found "one family described enjoying a newfound sense of cohesion and camaraderie with their neighbors." However, she says, "they experienced conflicts and tensions because of the differences in what people needed and the fact that their reactions were sometimes poles apart."

"Part of the issue was that the impact the fires had on homes was sometimes vastly different; some were still standing and others right next door were burnt to the ground." She notes that acute traumas can also lead to a whole set of other consequences that include having to relocate, the expenses of moving and potential time off work, and more.

Kia-Keating says research shows that traumatic events like disasters can shake loose people's sense of basic safety. She adds that the "intensity of personal and material losses can lead to complete and utter disorientation. Some people I worked with described a period of time when they just couldn't focus on anything, walking around with glazed expressions, not recognizing familiar people and places."

Even after that initial phase, people can experience a cascade of challenges, particularly in their connections with others. They may begin to feel isolated because they are still struggling and feel like they should be "over it," she says. "Several families I spoke with after losing their homes to a wildfire just fell apart over time. In the case of a husband and wife, they coped with the disaster very differently and found they didn't know where their common ground had gone or how to find it again."

At the same time, she says, "there are positive outcomes that people and communities can experience, including finding new meaning and increasing their altruistic and prosocial behaviors toward others. This kind of resilience, or even what's called post-traumatic growth, can happen for anyone. But no one should feel flawed in how they respond. It takes time to build the capacity for resilience, and it's especially helpful to take a trauma-informed[12] lens to help support people and communities in developing it rather than judging those who don't show it right away."

Culturally Accountable Trauma Response

It is important to acknowledge that trauma and the way it is experienced is not monolithic, and therefore how we communicate in traumatic situations must be culturally responsive. Indeed, trauma can differ across and within individuals, communities, cultures, and events, as Kia-Keating's work shows.

Even the word *trauma* can be difficult. "People were living complex lives before we called it trauma," says Theopia Jackson, the program director for Humanistic and Clinical Psychology at Saybrook University from chapter 4, who is also an accomplished scholar-practitioner and educator specializing in cultural accountability. "The language that we use is constantly evolving and therefore slippery." For now, she prefers to talk about complex trauma.

"Part of the challenge with the 'trauma bucket' is that it's become a catchall phrase that means different things to different people. It can lead people to think of completely opposite things, either a permanent state of being traumatized or as a thing that happened one time and is over now." Jackson notes that people also tend to think of trauma as something tangible, when a traumatic event may not even register as such to the person experiencing it. She says, echoing Kia-Keating, that complex trauma might help to think about how different people experience the same disaster.

"If someone has lived a typical life where they've had some ups and downs but they've been able to bounce back, enduring a fire or the loss of a home is one kind of experience. But for someone who's been bombarded with other significant challenges, like the death of a loved one to community violence, by the time they lose a home, they may have a very different response. And then, based on someone's social context, the community responds differently to their loss. Oddly, resources tend to go to the already more well resourced," says Jackson, harking back to Diamond Holloman's hurricane recovery experiences in North Carolina described in chapter 1.

However, Jackson says, it's important not to position people who are dealing with significant issues as having something less about them. Some folks are quite adept and functional despite the context they live in. Their world is often already much more in perspective. So when that fire happens, they may already be closer to the realization of "Hey,

I'm still here living," as opposed to their counterpart who has not been exposed to some of the worst things in life and may then lose their footing completely when faced with a similar situation. It's not about pitting one against the other but about showing the complexities for both.

Jackson notes that many communities of color, including socioeconomically diverse communities, have their own "network of care" that they're practicing in plain sight but out of view of the dominant culture. Professionals don't always see these networks, because we have essentially said to them, "You have to come into my office, during a time that works for me, to get to the resources I have to give you." But there's a whole other network of care happening outside of my office after hours. So what if instead, we could go meet them and say, "Who's already helping? And how do we help them help more?"

The tendency, Jackson says, is the opposite. "Professionals can sometimes think that communities are not accepting what we have to offer, maybe because we don't perceive them as being as emotionally upset as we want them to be." Instead, she says, "the responsibility is on me as the giver to say, 'What do I need to do to help you be shored up enough to accept what I have to offer?' It's a much more relational approach, and it relies on cultural humility. To ask, 'How can I enter into your world to accept you versus you having to come out of your world into mine to accept my gifts so that I feel good about it?' The question for me is, 'How can I lighten your load and not add to it?'"

When it comes to climate change, fires, and other disasters, Jackson says:

> I try to learn from communities that have survived their own apocalypse. We have many communities that have done that, from Native Americans to African Americans to our brown brothers and sisters across the world currently facing huge challenges, and their apocalypse looks different. So how have they continued

to live their lives, show up, while thinking, "I may not come home tonight because of the color my skin." How do they learn to live with that threat? And what can people who haven't had that experience learn from it? And it's not to say that the groups I'm talking about have not necessarily been living their lives in a state of fear. It's more like, "I have to do this today, to live today, knowing that tomorrow is not promised to me. And what can I do today so that tomorrow is promised to my next generation?"

The term *vicarious trauma* is common—it's the kind of trauma that can come secondhand through other's experiences. Jackson also speaks of "vicarious resilience," or resilience that can be experienced indirectly. The possible contagion of resilience furthers Jackson's point that trauma is complicated and that it's valuable to return again and again to a relational view—one that does not hinge on a static positioning of the helper and the helped, the hero and the victim. As always, it's much more complicated than that.

Trauma and Healing

"When it comes to fire, there can be a sense that Indigenous knowledge is a relic of the past. This is not the case—Indigenous fire practice is alive and well,"[13] says Don Hankins. "Tribal knowledge and experience are frequently marginalized or devalued in environmental management, and relationships between managers and Tribes are often nonreciprocal. But Indigenous knowledge brought us through the major climate events of the past and is absolutely relevant to what we face today."

Hankins is a professor of geography and planning at Chico State and a Miwkoʔ, or Plains Miwok, traditional cultural practitioner. He has spent his academic career working on fire and water in California, with a focus on applied traditional Indigenous stewardship. In practice, that means that he teaches and does research, including fieldwork at the nearby Big

Chico Creek Ecological Reserve, where students and others have the time and space to learn to work with fire.

"I strive to be in the field because that's where I can do what I'm most interested in: applying Indigenous techniques to management and using scientific methods to assess the results. This work provides me opportunities to advance science, but also to keep a cultural lens on the landscape to assess the condition of, and changes to, traditional resources and interspecies relationships in the places I work."

Almost daily, Hankins travels to Chico, then up through the foothills of the Sierra to the reserve, and then even farther uphill to his own house, which is just above the town of Paradise—the one that burned in the Camp Fire of 2018, the deadliest in California's recorded history. The areas where Hankins has been able to burn—namely the reserve and his own property—are calm, quiet, and though the faint smell of smoke lingers after a burn, it's pleasant, earthy. It's a stark contrast with the center of town in Paradise, where the chemical smell of burnt cars and tires and numerous other substances lingered long after the fire was out.

Although Hankins had to evacuate his own home due to the Camp Fire, it remained intact, and he doesn't harbor the same fear that many of the state's residents do these days. He's been working with fire for a long time. He introduced his kids to it early—his daughters have been around burns since they were toddlers.

It's not hard to imagine the world that Hankins does, where people are comfortable with fire in the same way that he is because they are able to develop a better relationship with it in the same way he and his daughters have. "If we teach people to read the land, they will know it is a drought and when good fire can be used instead of being vulnerable to what nature will provide otherwise. If we are to succeed in living in this land, we must consider what it is telling us and not force unrealistic solutions on it."

However, he knows that effort is not without its obstacles. "Integrating Indigenous knowledge and people into ongoing management efforts can preserve traditional ways and invigorate agency approaches, but my own research has shown it can also subjugate Indigenous perspectives," says Hankins. His work with Australian geographer Christine Eriksen of the University of Wollongong documented some of the difficulties faced by Indigenous people in both the United States and Australia as they attempted to navigate working for the fire service.[14] Most found they were unable to use their cultural knowledge in firefighting and that it was in fact frequently outright dismissed.

"There is a great deal of resistance regarding how traditional cultural burning can be carried out in partnership with agency-based programs. Rather than recognize the knowledge and preparation that traditional cultural practitioners have, agencies see their standards-based method as the only path to putting fire on the ground. This in turn risks traditional knowledge of fire and related cultural practices. Burning is a traditional sovereign right, but in many places, including the United States and Australia, legal systems work to regulate fire out of the land."

Hankins says "one way forward is to embrace traditional Indigenous law, which is rooted in nature and holds individuals accountable for their actions in a reciprocal relationship with the environment. I see using traditional knowledge and science in combination as being the solution to sustainable living and a resilient future. Really, that has been proven by the existence of the landscape that non-Indigenous people found when they arrived in California."

Whose Trauma

As Theopia Jackson points out, the very nature of the word *trauma* can be a challenge. Particularly when it is used to describe people who would perhaps not apply it to themselves. Patricia Fifita is a postdoctoral research fellow with the University of Hawai'i at Mānoa's Cooperative Extension program and a Pacific Islander from Tonga. She has been working to support climate literacy, adaptation, and resilience in Hawaii and the United States Asian Pacific Islands for several years.

During a presentation to a group of academics and practitioners from around the islands, Fifita outlined a diversity of Pacific Islanders' concerns related to climate change. "I think it is critical to look at the various narratives that frame the way climate vulnerability is understood in the Pacific and the abundance of meanings that are associated with it that do not necessarily serve to support or empower the communities living at the front lines of climate change," she said.

She gave an example of how small island nations are regularly held up as victims of climate change without being major contributors to the problem. Yet Fifita said:

> It is most often not the people that live in these places that are pointing to themselves as traumatized, though they are concerned. Instead, it is major environmental nonprofits and others that can exploit those stories for conservation gains that may not in any way return to the people whose stories are being used. Fortunately, there are some efforts being made to shift to a more empowering narrative that focuses on climate resiliency as opposed to climate vulnerability. This is not only empowering but also acknowledges island peoples' long-held knowledge and embodied practice of caring for and managing their island environments.

Amy Cardinal Christianson, a Métis social scientist who hosts the *Good Fire* podcast and works on fire as it relates to Indigenous peoples with the Canadian Forest Service in Alberta, shares a similarly difficult experience. In 2011, the Slave Lake Fire forced the evacuation of the town's seven thousand residents. In addition, eleven Indigenous communities were also affected and forced to evacuate.

"Many of the First Nations people did not want to evacuate. They felt they had more than enough knowledge about fire to defend their own communities. However, there is a lot of grey area in the legal relationship between the province and the territories. In British Columbia during the 2017 wildfires, Indigenous peoples were being threatened with having their children separated if they didn't evacuate. That is something that obviously brings up a lot of historical trauma," says Christianson.[15]

"For the folks that did have to leave, they were given cash to offset the costs of evacuating, and suddenly a narrative started to appear that they were spending it drinking in bars in town. In other words, the 'help' being offered—from the evacuation itself to the money—was no help at all. Staying may have been better for most except the highly vulnerable, who were already being aided from within the community." She notes that many people did end up staying, regardless of the risk, and felt the damage to their communities was lessened because of their firefighting efforts.

Christianson and colleagues have gone on to create materials to support First Nations during evacuations,[16] advocating for everything from allowing those who can to stay behind to helping those who need to leave to shelter with other nearby First Nations to offering opportunities for ceremony. "It's hard for some people to see, but First Nations can even be more vulnerable because the kind of help they are being offered is not the kind they need. Many communities are actually resilient to wildfire because of their local, cultural knowledge."

Anthropologist Yanna Lambrinidou says misassigned narratives about trauma have also been present in the Flint, Michigan, water contamination case. She says that there, narratives of trauma caused to residents by contaminated tap water have been turned back on the community at times.

"Trauma narratives have been weaponized against residents to dismiss their unanswered questions about the water by suggesting that trauma has led to collective distrust of science and that this distrust has rendered people incapable of recognizing that the problem with their water has in fact been fixed. This claim overlooks people's persistent experiences of illness and carries with it the implication that those who are not satisfied with existing 'solutions' are incompetent, irrational, or overly emotional," says Lambrinidou.

"But when one looks at the complexities, uncertainties, and unknowns surrounding Flint's water system and the city's recent violation of federal requirements, it makes all the sense in the world that some residents retain a healthy skepticism about the quality of their water and about official assurances about its safety."

As these many narratives indicate, trauma is not a simple topic, and it is valuable for science communication practitioners to better understand the many ways that both trauma and healing impact communication and engagement efforts. Part of that work is understanding practitioners' own trauma and healing and how they affect their own work.

Where Practitioner and Community Trauma Intersect

As Theopia Jackson said, trauma is complex, and navigating the terrain of trauma is difficult for everyone. It can also be easier to see other people's trauma than to recognize one's own, particularly in tense situations where much is at stake. However, trauma isn't something that belongs only to others. In fact, taking relational work seriously means that focusing on the ways that practitioners are "in the work too"—not standing outside of it with all the answers—is crucial. Without that recognition, science communication practitioners can run the risk of causing harm, and not just in disaster situations.

When it comes to scientists and science communicators in particular, who are rarely trained to work with trauma, Kia-Keating offers some advice based on her own experience after the Thomas Fire in 2017, which was followed by a debris flow that killed twenty-six people. She says that "in the aftermath, there was a surge of connection between scientists and the community. People were extremely anxious to understand the science and engage with researchers to comprehend not only what happened but the potential dangers ahead."

It's valuable to think about disaster response in phases, Kia-Keating indicated. Most immediately, of course, information related to basic safety and return to routine is important. She goes on to say that "in the longer term, of course scientific information can and should be shared, but it's good to continue to be thoughtful and intentional about how and when to disseminate that knowledge to support community change and resilience."

She says recent events have opened her eyes to "the personal distress that natural science researchers and professionals have been experiencing, in part because they are so attuned to the potential risks in our environment. This knowledge can heighten their fear and sense of urgency. Those who help communities respond to disaster can experience secondary trauma, or vicarious trauma, where people who are responding to disasters or are otherwise exposed to traumatic material can experience the same kinds of post-traumatic stress symptoms."

The challenge, Kia-Keating says, is that while one solution for working with trauma is to reduce exposure to traumatic material, avoidance is not always possible for scientists and communicators. "They can't just turn it off because it comes with the territory. But they would be wise to attend to the role of trauma in their work, to minimize its potential negative consequences. One repository of evidence-informed tools is the Vicarious Trauma Toolkit,[17] which is free and easy to access."

She says scientists can also learn from journalists who have given thought about how to report from communities that have experienced disasters and other tragedies. "We do a lot to encourage people and communities to have their emergency preparedness plans and kits ready but not enough to equip themselves with psychosocial 'resilience readiness' in the face of disasters. I'm hopeful that as we raise awareness about these issues across disciplines that we'll see more energy and commitment toward trauma-informed practices and preparations."

Lizzie Johnson, the journalist from chapter 5, says she's still figuring out the best ways to work with her own emotions around reporting. Because she's frequently meeting with people during some of the worst moments of their lives in her reporting on California wildfires, she's careful to ensure that the people she's interviewing feel comfortable, but that's just one piece of it. "I get these reminders of stories all the time—I'll be looking through the photos on my phone, and it'll be smiling friend, mom, trees, and then boom, fire and death. It's jarring, to say the least. And you know, once the story has run, it's not like it suddenly goes away for you. You spend that much time with people and get to know their lives, so it's only inevitable that you care about them, and their stories stick with you."

Therefore, she says, even the people who are doing the listening need someone who listens to them. That means a therapist or a friend who will hold space for you so that you can talk about what you need to talk about without them encroaching on that space. I think something that I found, for me personally, is it can be hard to talk about it. And some people try to relate by telling you their own stories, but honestly, in some moments, I just need to be able to talk about my experience without any input. Having a place to share your own story is the best thing I can do to care for myself, which in turn helps others.

Journalist Wudan Yan says, "Science reporting is seen to be this very fact-driven thing. But what I've learned over the years is how all these things tie together. It's so much more than science; it's also how it connects with environmental justice and human rights. It's hard to ignore the emotional component of science reporting." Individual trauma compounds the difficulty of the material as well, Yan says.

"Sometimes I have to step away from particularly hard interviews that touch into things that trigger me in ways I may not even understand. Once, I was interviewing a young woman in a refugee camp in Bangladesh who was around my age. She had been kidnapped and forced into marriage and then had her child kidnapped by the father and was in the camp trying to get her baby back. That interview sort of destroyed me, and as soon as the photographer I was working with came in, I just had to leave for a bit to gather myself."

Her experiences led her to write an article with resources for journalists contending with the emotional toll of their reporting.[18] Her own coping mechanisms include working with others whenever possible to make the work less lonely, and when she can't, making sure she can see friends and take time every day to journal so that she can process her own feelings.

Therapists are also skilled at working with what triggers their own reactions. Jackson says:

> I might be the therapist working with you. But I still want to be curious about my own trauma history. And also, my work with you around your trauma exposes me to a certain level of trauma. So it can inadvertently add to me feeling bad for you versus empathizing with you. And sometimes, people who are helping, they are helping out of their own desire to help someone, as opposed to being able to see what that person needs. And then they get their feelings hurt when someone doesn't accept the beautiful thing that we are bringing to them. These are the kinds of things that can come up if professionals aren't aware of their own trauma.

On top of that, there can be a lot of pressure for people who are seen as experts to have answers, to provide solutions. Jackson says she thinks of a recent deadly police shooting of an unarmed Black man and how she and others went about facilitating the emergent wisdom of the community. "I was part of the group that went in to help just be with the Black communities. Of course, they want answers, and I say to them, 'We may not be able to come up with the answers. Instead, how do we prepare ourselves for the realization that it's likely no one's going to be held accountable?' So the focus of the conversation now is not about what I think."

Jackson continues:

> We start instead to focus on questions like how do you still let your child out of the house safely at night? I will always say, if people have the capacity to live their lives, we have to have the courage to witness their lives. As professionals, it goes back to cultural humility and valuing what people have already done and joining with them in solidarity. I'm there to support people so they can be in a different relation to the trauma, not to fix it. So when the thing that you know is going to happen, happens, they're ready.

This is the context in which we practice: lives have been deeply shaken in ways that are not necessarily visible. There is no doubt that we inhabit an incredibly changing world—economically, environmentally, demographically, relationally. People are experiencing racism, sexism, job insecurity, moves, illnesses, divorces, and other psychically manifesting heartbreaks, wars. In the following section, we will explore how to create places where practitioners can thrive and stay in this work for the long haul.

Takeaways
- Trauma is an increasingly pertinent issue for science communicators working in embodied circumstances, particularly but not exclusively in disaster situations.
- There is a rich literature and practice on trauma-informed work in many fields that is applicable in science communication.
- Trauma is not experienced in the same way by everyone; practitioners must take care not to impose narratives of trauma on anyone.
- In a cultural context, it is important to interrogate narratives of trauma and determine who benefits from them.
- Practitioners are invited to understand their own trauma and how it affects and is affected by that of others.

Questions for Exploration
- What issues did reading this chapter raise for you?
- Where do you see trauma—your own and other people's—in your work?
- Where do you see healing—your own or other people's—in your work?
- Whose trauma are you focusing on?
- What would trauma-aware or trauma-informed perspective look like in your work?

PART III

The Future of Science Communication

CHAPTER 8

Equitable, Inclusive, and Just Science Communication

Until recently, science communication advice was seemingly agnostic as to who the practitioner was, although the implicit assumption has been largely white, male, with tenure at an elite institution. Simultaneously, many science communicators spoke to a mythological "general public," in which everyone was lumped together. It was assumed the same strategies would work for all—practitioners and communities alike—and that factors such as race, gender, sexuality, age, ability, and class did not affect the communication and engagement process, much less power and authority. Even today, discussions about the diverse people who are doing science communication work and why that matters are held at the edges of the field.[1] While marginalized, these are hardly marginal matters. In fact, they are central because who creates and disseminates knowledge, including the languages they use,[2] affects not only what is considered valid but also who influences the questions that are asked and benefits from resulting knowledge.

As a recent example, thirty-five female scientists wrote about working on the coronavirus and navigating patriarchy at the same time. "Neither epidemiology nor medicine are male-dominated fields, but women are quoted less often—sometimes not at all—in articles. What's more, the lack of inclusion of leaders of color is striking and disenfranchising for minority women scientists of color, particularly as communities of color are being hit hardest by this epidemic."[3] Although these researchers and science communicators were speaking about COVID-19, the issues they raised are the same ones that appear over and over again in science communication practice.

Rebekah Fenton, from chapters 4 and 5, further explains how these kinds of gendered and racial dynamics affect her as a Black professional working in the midst of both the coronavirus pandemic and widespread protests in the wake of police brutality. "After the first week of the protests, I was in church. When my pastor started to talk about the current protests in the context of this very long struggle, I started to cry. That release was key for me, knowing I was going into a workweek where I'd be alongside other Black and brown medical students who serve predominantly Black and brown patients," she said in an interview.[4] "Systemic racism prevents us as doctors from promoting people's health and well-being in a broad sense, beyond addressing disease. I hear a lot of doctors say, 'Well I'm just going to take good care of my patients,' in response to the protests, and leave it at that. But taking good care of your patients means acknowledging racism."

To be sure, a failure to acknowledge the identities, knowledges, and lived experiences of diverse science communicators affects not just their careers but also the communities they work with. "We live in an era of abundant scientific information, yet access to information and to opportunities for substantive public engagement with the processes and outcomes of science are still inequitably distributed," wrote the authors of a paper on inclusive science communication, led by Katherine Canfield, after the first national conference on inclusive science communication.[5]

Because broad discussions of inclusivity are relatively new in science communication, it is also a dynamic area of conversation where even the terminology is contested. It can be seen through any number of possible lenses: diversity, equity, and inclusion (DEI) being one—with justice sometimes added to create the acronym JEDI—representation and belonging another. *Positionality* is yet another term that arose in geography as a way to acknowledge where one comes from and how they know what they know, giving a way to think about and work with power differentials, access, and gatekeeping. Discussions about antiracist practices are becoming more common and commonly debated as I write this text. To be clear, these are not interchangeable terms, and each has its own rich scholarship and practice. Interrogating and contesting their varied meanings is a form of negotiating viable and just science communication practices.

Before going further, for this chapter in particular, it is important for me to identify myself and the position that I am writing from. To restate a brief version of what I did in the preface, I am a cisgendered white woman with a doctoral degree and a non-tenure-track academic position. For some, these may seem like superfluous details, but I state them to indicate that these factors, and many more that I am unlikely ever to share widely or even recognize, inform my outlook on my science communication practice and affect whom I am able to connect with. In coming to understand this, I am grateful for the world-changing work being done by a brilliant group of practitioners with diverse experiences and expertise. Instead of layering my own necessarily limited point of view on top of the perspectives they have generously described, I am sharing their work and their words.

This chapter is an invitation to science communicators to approach people as individuals and communities as unique with an eye toward reducing potential harms for all. The people and initiatives in this chapter demonstrate why relating—rather than theorizing—is vital to releasing the field from communicating in abstractions.

Changing Behavior Is a First Step

"For me, as a queer person, a trans person, I don't need my coworkers to agree with me; I just need them to behave in certain ways. It's that simple," says Cian Dawson, a hydrologist with a government agency. "Behavior is something that can change more quickly than deep beliefs, and 'random acts of reform' can change the overall environment eventually."

Dawson relays a concrete example of this kind of behavior change in a science communication context:

> When medical professionals change their language to talk about bodies and not genders, it's immediately seen as more inclusive by the trans community. These are small, uncomplicated language changes, using terms like *parents, pregnant people*, things that are not tied to male or female identities. People say we have to wholesale change everything, but I can't renovate the whole house at once—it's not livable. You have to work on multiple scales at once.
>
> From a simple utilitarian perspective in science communication, we need more people with a wide array of backgrounds who can actually bring different insights and reach different communities.

Dawson says, "We can't pretend that we can get away without finding a way to work better, affect change, and support communities that have been differentially impacted by things like climate change or drinking water and air quality. Science communication is always done by people, and science communicators' experiences are not indistinguishable."

Dawson's view is informed by having worked in many different sectors and seeing these kinds of tense communication patterns up close. "I spent a decade in science education, working in museums and doing public programming. I worked in schools, and I ran professional development programs for teachers," says Dawson. "Then I went back to grad school in geophysics—I started a PhD program but stopped with my master's degree, mostly due to diversity issues. I've worked in the government and in the nonprofit sector with a local environmental justice watchdog group. In my personal life, I've done outreach and education on queer and trans issues." Dawson is frustrated by the lack of progress on inclusivity.

"Every institution is now claiming to be doing diversity, equity, and inclusion work, and most of the time it just makes me angry," says Dawson.

> We had an institution-wide webinar focusing on inclusion, and the main message was that we should just talk to each other when we have a problem. That is a common means of dismissing complaints about harassment and aggression, about microaggressions and discrimination. People can say, "You're just too sensitive. I didn't mean it." But in the end, *I don't really care what you mean; I care what you do.*
>
> After that training, my comment was, "I can no longer recommend this agency as a workplace." To me, a modern workplace is one with the best opportunities to do good science, and that means a fully functioning organization that does not allow for harassment, that values inclusivity. I go back to the idea of creating systems that mandate behavior because you can train people all you want, but there's no teeth.

Dawson says, "If my only recourse to harassment is filing a formal complaint, I might as well resign because that's going to destroy my relationships. And it may not affect the change I want. What I want is to be in an environment where I can focus on my work."

These are not abstract issues for Dawson. They affect his sense of belonging and safety:

> I was doing fieldwork in a very remote area, with people I didn't know. I made a comment about my family and someone literally said, "I didn't know you had a family." I was blown away, like, I didn't just pop out. It was a telling moment because within that same conversation, I knew one person was engaged, another had a new job and awesome kids. But with me, it turned into a "don't ask, don't tell" thing where people don't even see me as a whole person. Even now, I work for people who I have been out to for years, but I can't approach them about challenges that being trans brings up for me in the workplace.

Dawson relays another experience from work at a remote field site, saying:

> I took a plane ride in and out, with people I didn't know. We were all sharing rooms, and when I say "sharing," I mean beds a couple feet apart with people I'd never met. There weren't toilets. It's the tundra, and there's no privacy and twenty-four hours of daylight. I'd wander off as far as I could, but there was often a drone flying around to capture field images. I couldn't even figure out how to troubleshoot—could I ask for my own space? And it affected my health because I could rarely go to the bathroom, and I didn't fully change clothes because I need about twenty minutes alone to do that and never had it.

Those kinds of experiences have left Dawson certain about the reforms that are needed, saying, "In order to succeed in science and science communication, you have to demonstrate your ability to adapt and learn new information, new jargon, or changes to conceptual models. If you can do that, you can respect a person's name and pronouns. It's that simple. We can't continue on unless we build better institutions where more people can be safely involved."

But how we get there is still fraught territory, Dawson notes:

> A lot of diversity conversations don't have teeth, because even who gets chosen to speak is laden with meaning. People don't want to hear somebody who's angry, so we never hear that. At big conferences we have panels, and nobody new comes. Panels rarely address the hardest topics. There's no opportunity for conversation. But then when you do mandatory training, it just becomes a check mark and still doesn't reach the people that most need it and is traumatizing for others. It feels like we're in this tense space where we see all the problems but none of the remedies is yet effective.

Developing Trusting Relationships
"I was invited to the Grand Canyon for the hundred-year anniversary of the park, but there was no acknowledgment that the land was there long before, that there were people there long before," says Sergio Avila, a scientist and outdoors coordinator with the Sierra Club in the southwestern United States.

"Viewing humans apart from nature is a myth and a very European invention. There is a history of racism in the field of conservation that has resulted in scientists being comfortable talking about 'prehuman' time in places where humans have always been," he adds. "As part of communicating science, we need to break the paradigm that environmentalism and conservation are separate from social justice."

Born in Mexico City and raised in Zacatecas, Mexico, Avila says he is following in the path of his parents, both of whom are scientists, though he is taking a slight turn from their careers as professors. "I've had a lot of access, privilege, and comfort with learning and navigating academia.

As I started out, I could easily see myself on that traditional path to success—many publications, a big lab group. Instead, I worked with Indigenous people and found a different calling. I realized that for me, the path I was on was not going to give communities what they needed. They had immediate needs I could help with, but it would have to be in a different role."

Avila was researching endangered jaguars along the border between the United States and Mexico. In addition to working with Indigenous peoples, he had to learn to engage with numerous people with distinct lived experiences. Simply finding jaguars and gaining access to their territories for data collection was difficult. To do that, he had to work with landowners and quickly saw that "there was no instruction for this part."

"I had to explain my work to landowners and realized all the tools I had to do that were exclusionary," Avila says. "Too often, scientists create language that is actually meant to exclude people. We build ourselves up as the experts. And then we're supposed to 'dumb down' or simplify our language, which just leads to more feelings of superiority. Scientists walk into a place they don't know at all, acting like they do know it all. In fact, it is very humbling to meet landowners who truly know a place in a way I never will."

Avila says that unfortunately there are still a lot of bad research and communication practices that persist. "It makes the people we need to work with skeptical. And that's hard because we scientists, especially people of color, have to do a lot of emotional labor."

Emotional labor is the process of managing one's own feelings and expressions to fulfill the requirements and expectations of a job, which is enforced by the dominant culture. This functions to keep others happy and is largely invisible, unrecognized, and unpaid, which results in a mental load or burden not carried by everyone, though it can also be explicitly taken up to help "share the load" as well.

"We scientists and science communicators don't always see that we are asking communities to do more than their fair share of emotional labor too. It can lead to community burnout." Science is a tool, but it can't be used to exclude people, notes Avila. "The idea of 'science literacy' is not inclusive. We ignore so much—language, learning outside of classrooms, eating meals together."

Avila learned that efforts like shared time are crucial, but sometimes he had to learn the hard way. "I can be pretty funny and I tend to talk a lot, and while it can be useful sometimes, I've also been told that I need to tone it down at times. Those community interactions lay bare who you are as a person, aside from the degrees and institutional affiliations.

"I've had to learn to listen more, to offer less 'solutions' and advice that nobody wants. You have to talk about things you have in common and see where things go from there. I have a lot of ranchers in my family, so I was able to start there to create some trust with ranchers whose land I was trying to access."

Trust is also key for Avila:

> In Mexico, there is virtually no public land, unlike in the United States, and so the dynamics when it comes to research access are also quite different. Ranchers and landowners on both sides of the border have their own community and talk a lot; therefore, they tend to view predators like jaguars in similar ways. Mexican landowners might fear their lands being taken away if they disclose there are jaguars, not necessarily seeing that in the United States many people are working on public lands with grazing allotments that have to be renewed. Developing trusting relationships, following through, is the only way to do this work.

Letting Science Come Second

"I love liminal spaces, in-between spaces, edges," says Joanna Nelson, an ecologist and boundary-spanning professional working primarily on the central coast of California. Nelson focuses on something that is commonly divided: the interaction between land and sea. As most ecologists know, there are terrestrial people and there are ocean people, and it's rare that the two collide professionally.

Even her research and consulting company, LandSea Science, was born out of a desire to create an institutional home that didn't force her into one niche or the other. It also allowed her to continue to work in the way that she wants to: in a holistic way on land and sea interactions, with a focus on relationships, care, equity and inclusivity, and justice.

In practice, that means that Nelson works on a wide variety of topics and in ways that can fall between the cracks. She has been working with the Amah Mutsun Tribal Band and their land trust efforts to adapt to climate change and restore their lands. In that role, and as a white woman, she follows the lead of her Indigenous collaborators and is there to support their efforts as a guest on their lands. "That could mean anything from, as Wendy Smythe says, aligning and braiding knowledge between Western and traditional science to physical labor done with the Native Stewards to reduce fuel loads and restore native, perennial grasslands."

When it comes to science communication and working with frontline communities, Nelson says she begins her work by asking what is needed, where she might be able to help. "If I can help through my science training and conservation experience, great, but I don't assume that what's needed, all the time, is Western science. That's usually not the gap. I can show up with what is needed, if possible, and then refine my awareness about when my voice can add to movements and when it's time to step back in a supporter or follower role."

Nelson is also involved in gender justice efforts in STEM. With training from Gender Spectrum, an organization that provides gender-affirming care and education, she works with schools and her own Wilderness First Responder community, of which she has been a longtime member. In addition, with colleagues, Nelson is working to establish a series of workshops that focus on embodied storytelling and improvisational approaches to connect and support women, femme, and nonbinary people so that they can thrive in STEM fields that can be hostile to them.

In an interview with Promita Chakraborty, Nelson acknowledged the many difficulties faced by women of color, people of color, LGBTQIA+ folks, and white women in STEM, as well as those who are no longer in the field because they needed to leave. As an antidote, she says, "We envision a collaborative, interconnected community—or network of communities—grounded in compassion, listening, and the understanding that we will succeed together through embracing equity and our shared humanity. It's a culture of mentorship and collaboration."[6]

Part of that vision, Nelson says, means acknowledging different ways of knowing and the knowledge that comes from lived experience, as well as valuing relationship skills and emotional intelligence. "Being told you don't belong, implicitly or explicitly, is wrong. We can change the culture of STEM when people can work without additional burden, distress, disease, assault, or other forms of harm. A new culture of STEM would reflect the truth that we're in this together and my liberation is linked to yours."

Nelson is quick to acknowledge her privilege as a white woman with a doctoral degree, straight and married. However, the death by suicide of her sister, followed quickly by her father's death, as well as a roommate's death, all in her early twenties, led her to grapple with what she calls "big grief." She says these experiences have deepened her commitment to making STEM fields more compassionate, welcoming, and just. She says in this work, "our love for the world and the people in it is foundational."

Making Science Communication Accessible

"We, along with other scientists with disabilities and medical conditions, have found success in our careers, but only because we have had access to health care, emotional support, and institutional backing," Gabi Serrato

Marks and Skylar Bayer wrote in *Scientific American*.[7] You may remember Serrato Marks from chapter 5, where she offered advice on making some listening exercises more accessible. She studies climate change through the lens of precipitation patterns in Mexico as recorded by cave stalagmites. Serrato Marks also has Ehlers-Danlos syndrome (EDS)—a connective tissue disorder that makes it challenging to do many of the activities that graduate students in the geosciences do, including field and lab work and teaching.

As a result of what she has learned in navigating her own disability, Serrato Marks offers training on accessibility in the sciences. "People tend to think about ability and disability as separate states." She says, "In reality it's more of a spectrum. I define *disability*—and, yes, you should just use the word—as 'anything that impacts your ability to work, learn, or take care of yourself.' There are many types of disabilities, ranging from mobility to chronic illness to learning."

Serrato Marks is broadly committed to accessibility in the sciences and is particularly interested in how to make field courses and experiences more accessible, with steps like ensuring accessible transportation and visiting sites designed for access. She says that academia, and particularly her own geoscience field, still has a long way to go.

The lack of accessibility has a career cost for Serrato Marks, who says she cannot follow a traditional academic path to becoming a professor. She explains, "While academia can be flexible, which I need, with EDS I can't work seven days a week with so much pressure to get funding and publish. It's difficult to stand at a board and teach—right now, I have a student that writes things on the board for me so that I can teach while seated."

At the same time, she notes that EDS has helped her to become a better science communicator. Because it is an underdiagnosed syndrome, Serrato Marks had to find creative ways to explain it to experts and nonexperts alike. She also has advice for other science communicators.

"Inaccessible science communication is not open, equitable, or inclusive," she says. In workshops, she presents best practices by using image alternative text, which describes images and videos on social media to readers who have visual disabilities, providing information in multiple formats including audio and transcripts, and striving for accessibility versus compliance with regulations.

Serrato Marks also urges science communicators to ensure that if they are doing things like writing about disability not to do so in a way that is separate from people who are disabled. When you are communicating about a disability, she says, "reading as much as you can about disability, preferably written by people with that specific disability, is crucial. Spend the time to get to know people, to interact with them, to relate to them."

Although it can be difficult, Serrato Marks notes that she's found a supportive community in her work on accessibility. Like many others in this chapter, she says that community is foundational. "It's been so meaningful to meet other people that are on a similar path and facing related issues. I feel like I now have a community of people that I can talk to in terms of both commiserating and working to make things better. We can relate to each other for the long term, regardless of the research we are working on."

Healing and Building Together

"I was not somebody who was interested in improv. I was interested in people building things together," says Raquell Holmes, a cell biologist who has gone on to found the organization improv**science**. "I was, and am, interested in relationships, in how we live and work together. Helping people has always been at the heart of my work. I created improv**science** to help scientists bring to each other, and to the world, our compassion and love for what we do."

Like many graduate students in the sciences, Holmes had therapeutic support for getting through the multiyear process. She was introduced to a form of group-centered social support that she continued to practice over decades, becoming a coach herself. "I got involved in social therapy and in what is called the performance approach to human development. It's a non-diagnostic, group-centered, community-based approach to development. I was able to find an environment that supported my growth emotionally."

Holmes says these group-based approaches grew out of a collaboration between social workers, community organizers, and activists who were interested in a practice "that was emotionally helpful and an alternative to traditional therapy. There is a group leader who keeps everyone's attention on the group, helping it to continually transform how it's working together so that it's always oriented toward development. We're always challenging power because we continually ask, How are we building with one another?"

This social and group orientation grew out of Holmes's grassroots community organizing:

> I volunteered with the All Stars Project—an afterschool program using performance to support young people from poor communities—this is how I was introduced to improv. I was standing on street corners, organizing kids to perform. There were people from different backgrounds—class, race, and age—who didn't know each other but were working together, and we would use improv techniques to help them relate to each other. Too often, we label one another and end up not looking at the nuances, the crevices, in terms of what makes us unique. I want people to have relationships where we're interested in each other, where we're not attacking each other for who we are.

Holmes says that people around the world are using improv and other performance approaches to address social, societal, emotional, and humanitarian needs. "I saw this practice that I'd been involved with for a long time turn into something that executives and Fortune 500 companies were using to help their employees work better together, and I realized I could combine what I knew about group human development work and improv to help scientists create environments in which we're growing and building together."

"When I lead a workshop, I'm less interested in people learning improv than them being improvisational. I want them building with each other," Holmes says. "I want people to have a transformative experience and come away with a new understanding of their capacity for being present with others, for listening, for being changed or informed by what another human being does and says. Science communication at its best is a relational art form. You have nothing except for what's created from the dialogue with the other. What's important is that we are creative, that we are evolving in relation to one another."

For example, "In an improv scene, it's very clear that we're in this together. The question is what do we want to do with what comes up? My job is to keep the other person in the conversation with me. I don't want to deliver a point and walk away—I want to keep the conversation going. I think that's the most beautiful part of the practice, this bit about how to stay connected and how to keep building relationally. It's key to science communication practice."

Holmes says that the group-oriented improv practices she leads "create an environment in which we're able to be well and do no harm. Creating conditions together, in which people can choose to speak or not, builds collective well-being and action." Holmes's group-oriented practice is aimed at developing that collective, emotional well-being.

Laughing and Truth-Telling

"Toward the end of my doctoral training, I suffered from terrible depression, and it compelled me to leave graduate school," says Kyle Marian

Viterbo, a science communication practitioner and comedian. "Luckily, I had supportive mentors who helped me find my way toward a master's program in science communication at the University of Edinburgh."

Within the first couple weeks there, a classmate encouraged Viterbo to attend an event where scientists and other academics were doing stand-up comedy. "I was struck by the last person to go up—a young woman who had just defended her dissertation and spent her eight minutes pointing out all the weirdness inherent in that process. I was amazed to see the level of subversive truth-telling she could do in a comedy format." That night, Viterbo signed up to do her first show and hasn't looked back.

After returning to New York, she went on to develop *The Symposium: Academic Stand-Up*, a "show and workshop series that uses sharp, socially-mindful comedy to challenge academic norms and champion inclusive science communication." Her thematic shows have addressed subjects ranging from the perceived neutrality of museums to sex and gender norms to science and race.

Viterbo's show *Asians Strike Back: A Coronavirus Comedy and Science Show*, codeveloped with comedian Esther Chen, was timely. When it became apparent that COVID-19 was becoming a pandemic, they jumped into action to address the ways Asians were being targeted and harmed both verbally and physically because of the perception they were responsible for the virus. Viterbo and Chen described the effort as an "all-Asian comedy show exploring the science and social repercussions of racialized emerging diseases." The comedy was mixed with presentations by scientists talking about the virus itself, as well as public health measures like mask wearing.

Viterbo says the show had two primary goals. "The first was to demand space to spotlight the complexities of covert and overt anti-Asian racism, model-minority-ism, and what being a minority with white-adjacent privileges means. The second was to prepare audiences to fight what was an unknown, distant fear." That distance closed quickly, however.

Viterbo notes, "On the day we performed, New York City announced its first official COVID-19 case just as we were leaving the venue. By then, Chinatown businesses had already been suffering and deserted for at least a month." She says this is just one example that points to the need for more diverse science communicators. "Being attuned to sociocultural norms makes for better science communication—something that has literal life-or-death consequences in high-stakes settings," she says.

In a bit from her *#MuseumsAreNotNeutral* show, Viterbo offers a "critique of science's colonizer perspective through comedy" in part by reckoning with her own relationship with a childhood hero. "Indiana Jones ruined me. Indy was an inspiration for me growing up to become an anthropologist, a paleoanthropologist, to do archaeology, because he encapsulated the American dream for me: being a white man who didn't have to be good at his job, can fck off whenever he wanted, and never gets fired because he has tenure. Fck you, Indiana Jones."

After the show, Viterbo says she found support in unexpected places. "There were museum curators and educators in the audience. One walked up and said they knew they weren't doing enough, but this show—which featured stories from an Indigenous museum professional, Black comedians and historians, and others—spotlighted the impact of taking too long to make change despite folks in power having good intentions," she says. "There was also a table of older white women who were in town on a ladies' night out. They came up and very genuinely said they loved the show and that hearing these perspectives gave them a lot to think about. That it changed how they understand and appreciate what's missing in museums or how collections are framed."

Viterbo says that while the contexts and foci might change, "comedy has gifted me with the power to spotlight why making and demanding space for uncomfortable histories and topics in STEM matters. It's about justice and being able to train marginalized academics and science communicators in how it feels to demand space for diversity, equity, and inclusivity conversations." She says, "Comedy teaches a performer how to push the boundaries and test the limits of 'what's acceptable' for an audience. It can empower marginalized folks to break free from the pressures of prioritizing white and cisgendered approval and demand that inclusive spaces and conversations are prioritized too."

Representing, Belonging, and Justice
Black Birders Week was developed in a matter of days by a group of researchers, naturalists, and science communicators in response to a Black birder being threatened by a white woman in Central Park after he asked her to leash her dog in an area where leashes are required.[8] Luckily, the incident did not result in further physical violence toward the birder. It did further a national conversation about representation, belonging, and justice in the outdoors.[9]

That incident, combined with continued police violence directed at Black people, most recently George Floyd and Breonna Taylor, led a group known as BlackAFinSTEM to create Black Birders Week. The event took off to a degree that astounded even the organizers. "We expected it to be a conversation within the birding community. But we definitely did not anticipate the amount of overwhelming support—not just from individuals but also from entire organizations and even government agencies," said Corina Newsome, a graduate student in biology at Georgia Southern University and event co-organizer, in an interview.[10]

Building on the efforts by Stephani Page, who developed the original #BLACKandSTEM movement,[11] many participants spoke of finding a community larger than they had known existed. In response to a tweet from Texas teacher Baiyinah Abdullah that included photos of herself birding, Jameela Jafri wrote, "I've been a hijabi birder for over twenty years and this is the first time I've met a Black hijabi birder. I am so grateful to #BlackBirdersWeek for elevating Black voices and profiles. I just want to go birding with these amazing people. Right. Now."[12]

Alongside images and words of joy and solace, the organizers and participants shared their thoughts about addressing anti-Black racism. Earyn McGee is an event co-organizer and doctoral student at the University of Arizona, as well as a science communication practitioner who created the social media game *Find That Lizard*. She described some of the realities of being Black in the outdoors in an interview, saying, "It's always something that you have to have in the back of your head. I do try to enjoy myself when I'm out in nature. I have to be aware of the area that I'm going to and how I present myself. I need to make sure I have actual hiking boots, stuff that's labeled REI, and I come decked out so people know that I'm out here because I am interested in nature."[13]

"I think it's very important that people see and understand that racism is a direct threat to environmental progress. If you are an institution that prides itself on diversity, equity, and inclusion, but you don't address anti-Black racism or you don't address how, as an institution, you may perpetuate white supremacy, then I think your organization's longevity is at stake," said Tykee James, event co-organizer and the Government Affairs coordinator at the National Audubon Society, in an interview. "This is about changing the face of conservation and not just changing the topic."[14]

The success of Black Birders Week led to related efforts, including Black Botanists Week. Maya Allen, a doctoral student in biology at the University of New Mexico, organized the botanists' event along with several colleagues. Allen says the group chose to focus on connecting Black botanists—not just scientists but also artists and hobbyists—as well as educating. In addition, Allen says that founder Tanisha Williams, a postdoctoral researcher at Bucknell University, was able to reach a large group of international botanists based on her work abroad, leading to cross-continental connections.

Allen was particularly proud of Black Botanical Legacies day, which highlighted the historical contributions of Black botanists, including

Antoine, an enslaved man who created a popular new variety of thin-shelled pecan in the mid-1800s.[15] "Black people have had long-standing relationships with land. I hope that bringing some of these stories to the forefront will help people incorporate them in the classroom," she says. "I was happy to have people quickly message me saying they were going to do just that. It's been exciting because these are stories I didn't learn in my botanical education either. It's been great to develop this fellowship of Black botanists, to share plant stories and plant joy," she says.

In addition, Allen notes that she wanted to show "it's not an exception for Black people to pursue careers in botany. I hope that highlighting these amazing Black botanists normalizes these interests for younger people. I want Black students to see they belong and are not alone." However, Allen says that there can be other barriers for students. "Things like sports and medicine are seen as the best ways to achieve financial security. Unfortunately, many undergraduate research opportunities are unpaid, and students who can't afford to work for free don't always have ways to gain experience. Good mentorship can help. Personally, I get fulfillment from working to ensure the next generation will have a better experience, a more fluid journey, and is not going to face some of the discrimination I have."

The events have already had a big impact. Abbigail Turner, who will be starting graduate school at the University of Illinois, tweeted, "I'm excited to go to grad school but also nervous. The lack of diversity in STEM has led me to question my capabilities and my place in biology. #BlackBirdersWeek has truly reinvigorated and reassured me that I'm not alone. I can do this, & I BELONG HERE."[16]

Inclusive SciComm

"I went to a few national science communication conferences one year and walked away stunned by the fact that there was no real mention of inclusion or equity. Most of the speakers were white men, which was both exclusive and excluding," says Sunshine Menezes, a clinical associate professor of environmental communication at the University of Rhode Island (URI) and director of URI's Metcalf Institute. "Simultaneously, I was watching a community of science communicators grow on Twitter, with an entirely different set of people. Many were young, diverse, all over the country, working on so many different topics and in interesting ways; inclusivity was a given. Being in those separate worlds created cognitive dissonance, and that planted the seed of what became the Inclusive SciComm Symposium."

"We launched the first symposium planning committee with colleagues from across URI who are engaged in social justice work, including filmmaker and journalism professor Kendall Moore, biologist Bryan Dewsbury, anthropologist Amelia Moore, and neuroscientist Alycia Mosley Austin." Menezes says after some initial planning and input of funders, they broadened the organizing group to be more nationally representative:

> I'm grateful that colleagues including Mónica Feliú-Mójer of Ciencia Puerto Rico and iBiology and Shayle Matsuda from the University of Hawaii got involved, as well as Hollie Smith from the University of Oregon and Christine Reich from the Boston Museum of Science. A dedicated group of people pulled the first conference together in less than six months.[17]
>
> We wanted to model inclusion and be broad in how we defined the topics so people would see themselves reflected in the program. The participants included many people who don't talk with each other enough—librarians, youth educators, academics, and practitioners—and it's turned into a rich community that I'm excited to be a part of.

By the second day of the first conference, participants were already asking about another gathering. "We wanted to focus on equity and inclusion because it was so absent in other formal science communication spaces, and we immediately saw the demand is there. We could probably hold a couple of these a year and not meet the need."

For science communication practitioners, the meeting agendas have offered an explicit way to address issues that rarely reach the more normative science communication conference stages. Attendees had the chance to watch *Can We Talk?*, a powerful film by Kendall Moore that explores structural, cultural, psychological, and institutional obstacles that serve to limit the inclusion and participation of underrepresented people of color in STEM.[18] There was a broad offering of sessions ranging from critical disability studies led by Lisette Torres-Gerald of Nebraska Wesleyan University to recognizing and addressing privilege led by Catalina Martinez of the National Oceanic and Atmospheric Administration.

Although the conference has been successful, the work itself is still in evolution. Menezes says that even the name of the conference is an open discussion. "We have received pushback, especially from younger folks, about using the term *inclusive*. Truthfully, we did this on short notice and wanted a name that could get some traction. Is it the best term? No. In general, younger science communicators start from the assumption that communication must be inclusive and equitable," Menezes says.

> Therefore, calling something inclusive can be seen as undermining that foundation, which I understand. My response right now is that we need bridging language to connect these conversations that are all about inclusivity with those that ignore it.
>
> Once upon a time I was a taxonomist, and we talked about "lumpers and splitters." For the sake of this conversation, I'm a lumper: when I talk about science communication, I mean it in the broadest possible sense, as any interaction between people about STEMM (science, technology, engineering, math, and medicine). Hopefully that helps many people see themselves as

part of this conversation. But the language is tricky, and we're in a transitional period.

Menezes says, "We're working to push equity and inclusion to the center, to say it needs to be a part of everything we do. This is a significant shift from today's status quo, and this sea change is essential to achieve effective and meaningful science communication."

Beyond the gatherings, Menezes and her colleagues have been evaluating the inclusive science communication landscape and have results that are valuable for the field. "We've found that many graduate students and early-career folks are forging important new paths for science communication, but for the most part, institutions don't recognize or acknowledge their leadership. There are structural impediments to doing this work institutionally; that's just the truth. But these younger folks are ready to lead," says Menezes. "We need more established people who have job security, or other kinds of power and influence, to open doors and hold them open and bear some of the risk for younger leaders in inclusive science communication, because it is risky work. It is important for more senior people to give up some space and power. Even then, systemic change won't happen as quickly as many of us would like. But it can change, and the more people who are pushing, the faster it can go."

An Effort to Reclaim STEM for All

"Underrepresented people in science need unique resources to succeed, and only when institutions invest in these resources can science communities can finally be more inclusive. While many institutions are talking about diversity and inclusion, academic institutions must also invest in the resources necessary to retain STEM students from diverse backgrounds,"[19] says Evelyn Valdez-Ward, a doctoral student at the University of California, Irvine, who studies how drought affects the interactions between plants and their soil microbes. One place that she says marginalized scientists need better resources is in science communication training.

After attending several science communication training efforts, Valdez-Ward and her colleague Linh Anh Cat, now the chief of Resources Management and Science with the National Park Service at Cabrillo National Monument, came to the conclusion that neither the training nor the people in the room—trainers or attendees—reflected their experiences. So they came up with an effort of their own, which they call Reclaiming STEM.

"The workshop is the first of its kind created to center science communication and science policy training, with a social justice lens, specifically for marginalized scientists," they wrote.[20] The organizers include LGBTQ+, people of color, those who identify as women, disabled people, and first-generation students in that group.

"We believe the increasing diversity of scientists requires training that accounts for the challenges that underrepresented groups face. These range from acknowledging race, gender, sexual orientation, and other factors that influence our journey through scientific training, to addressing the fact you can have secure careers in science communication and science policy and won't have to rely on unpaid internships and training."

They take an inclusive approach to organizing and carrying out their workshops as well. From opening with an acknowledgment of the Indigenous land they are on to the diversity of the planning committee to working with locally owned catering services, the group is demonstrating how feasible inclusive practices are.

"We want to support people in bringing their identity, their culture, their full selves, to this work. We always have a diverse and authentic speaker lineup and ensure that the spaces we gather and our presentations are accessible," says Valdez-Ward.

Although still a young organization, its first workshop in 2018 demonstrated a large interest in more specialized science communication training. "Many science communication and science policy trainings offer guidance and tips for success that may not be feasible for those who may identify as underrepresented minorities, first generation, of different abilities, or LGBTQ+. The response to this workshop showed a need in the scientific community for a space to discuss diversity, inclusion, and advocacy in STEM."

For a second workshop, they added Rob Ulrich, a graduate student and president of Queers in STEM at UCLA, to their organizing group. The trio is now working to grow their organization and held a science communication workshop on the East Coast, co-organized with Tamara Marcus of the University of New Hampshire, and are planning future workshops in various locations, including virtually. A working group called Academic Secret Menu has spun out from that workshop and is focused on illuminating the hidden curriculum of scientific training, where people can discuss everything from the value of attending office hours to the ambiguous process of applying for graduate school.

Valdez-Ward is also turning the workshop into a research project. She says that because diversity is rare in science communication and policy, she is working to "highlight the importance of representation in these training spaces, as well as the value of diverse identities in creating safe learning environments."

Retaining Diverse Scholars through Community

Aradhna Tripati is an accomplished scientist who began college full-time at the age of twelve and has been on the faculty at the University of California, Los Angeles (UCLA), since 2009—the first woman of color in the departments she is associated with—and received tenure in 2014.[21] Her lab focuses on the role of the carbon cycle in a changing climate and past climate change impacts to understand current earth systems and plan for the future. She and the group have published prolifically, including in *Nature* and *Science*. During the last several years, she has turned her attention to creating more opportunities for students like her—those who faced a number of barriers.

Indeed, Tripati's upbringing is one unusual for her field. She says, "My parents are from Fiji with Indian ancestry. They immigrated and dealt with racism, incarceration, and homelessness in the United States. The scope of these issues for students who like me are people of color, as well as people of various gender identities and sexual orientations, has drawn me toward fostering success for people from diverse paths who have faced various forms of oppression."

To help ensure students from underrepresented groups can thrive in science and community engagement, Tripati developed the Center for Diverse Leadership in Science at UCLA. The inclusive program awards fellowships to not only high school, community college, and undergraduate and graduate students but postdoctoral scholars and faculty as well. This intergenerational approach is intentional, aimed at fostering a community that can recruit, retain, and support students at all stages of their education.

"There has been a big focus on recruiting underrepresented students into STEM fields, but what we've found over time is that students want to be here, and staying can be difficult because we don't put as many resources into retention, which just leads to further isolation for the few that are present. Some people even experience being the first and only person with their identity amongst their peers in a department," says Tripati.

"Whether it's support for career readiness, financial stability, or familial matters, we work to support students to stay in the field. In our program, we offer all fellows some form of financial support, they are part of diverse research and engagement teams, and they receive mentorship. Simultaneously, our faculty fellows are also being trained to better support students with inclusive mentoring, teaching, and leadership practices. That full-spectrum effort is crucial."

Developing community is also key, notes Tripati:

> STEM environments can be isolating, and incredibly exclusive, impacting people who feel or are different from their peers. We work to create a sense of belonging and build an inclusive community of scholars, researchers, communicators, and professionals who are empowered as leaders. My goal is that this will also spread across all of STEM and higher education institutions. Our ethos can also have a reach far beyond the university and into the organizations our fellows become part of as they go through their careers. Currently, the field of environmental science is almost 90 percent white and 70 percent male, and many other STEM fields have similarly low numbers. The pipeline is not being developed, and there are high attrition rates for underserved groups at every educational stage, sometimes as high as 50 percent.

The center has already trained 118 early-career fellows and upward of twenty faculty fellows. Early-career fellows can propose and then lead community-engagement programs that they are personally and professionally invested in. It is clear in hearing from students that it has been the opportunity of a lifetime. They speak to Tripati's warmth and welcoming, to the community that they develop, and to the support that they receive. And many have already gone on to prestigious positions in other organizations. Juan Lora was a postdoctoral researcher with Tripati who led an outreach program to bring water tanks for learning about atmospheric processes to K–12 classrooms. He recently joined the faculty at Yale University.

Another student, Alexandrea Arnold, who is Chickasaw, transferred from community college to UCLA and graduated with two degrees before starting her doctoral training with Tripati, studying how precipitation patterns in the Southwest have changed. "When Alex graduates, she will be one of fewer than two dozen Indigenous women in the United States with a doctorate in atmospheric sciences, and that representation is important," Tripati notes.

> In my mind, science is for everyone. From climate change to pandemics to inequality, the issues we face are interrelated and take collaboration to solve. I am motivated by social movements, where many individuals serve as agents of change by listening, learning, amplifying, and finding ways to connect with others. In the center, I encourage people to see themselves as both teachers and learners, as community-minded scientists, and to give equal weight to their programs of research and public engagement. Before we started, there was nothing else like this. The demand is unbelievable. And the return on investment is huge. With our model, our fellows practice the skills we need to build and strengthen the fabric of a civil society, where they will be the connective tissue.

Toward Inclusive and Transformational Science Communication
The kinds of experiences and initiatives described in this chapter clarify the ongoing reforms that are needed in science and science communication to make it more inclusive, equitable, and just. Hydrologist Cian Dawson explains, "Some of the things that would benefit me would benefit everyone, and I'm glad to see people talking about things like how to make fieldwork more inclusive. There are so many things like workplace flexibility, accessibility, different learning styles that make workplaces better for everyone. However, I think we have to honor a range of identities and experiences."

Dawson goes on to say, "I worry that in science, we're making the kinds of mistakes we made in mainstream gay politics by advocating for people that fit into boxes—married with 2.5 kids and a white picket fence—instead of advocating for freedom and inclusion for everyone. I think the challenge is that assimilation is easier, so we need to be having conversations about making science and science communication more inclusive for everybody and how that itself will change the field."

Change invites a broad range of efforts by individuals and direct financial support from institutions. Science communicators have to not only confront their own personal biases to make this happen but keep learning about other people's lived experiences, too, and this is particularly true for those from privileged backgrounds. Practitioners must also take on internal efforts to change organizational policies, structures, and practices in service of systemic change and multiple truths and ways of knowing and thinking. Institutions must support this work as well; it is not enough to let it fall to marginalized practitioners to do in their "spare time" and without resources.

Ultimately, Dawson says, people and institutions need to take a stance. "For things to change, there has to be a decision about the value of diverse science communicators and the inclusion of diverse communities. In addition to respecting the humanity of all involved, if you fundamentally believe diversity is necessary for successful science and for successful society, these are basic tenets."

Takeaways

- For much of the last few decades, science communication advice has not taken the identity or position of practitioners into account.
- This oversight has led to misguided training efforts and to the exclusion of many practitioners and communities.
- There are many practitioners whose work models the varied ways that science communication can become more equitable and inclusive, as well as questions whether that is, in fact, an appropriate framing.

- The personal narratives in this chapter encompass a broad array of approaches that practitioners are putting into action in what is a critical and quickly evolving component of science communication practice.
- Both individual and institutional efforts are needed to support inclusivity and equity in science communication.

Questions for Exploration
- What issues did reading this chapter raise for you?
- How do you describe your own positionality in your science communication work?
- How are your science communication practices inclusive and equitable, and where could they improve?
- With whom could you collaborate or work in solidarity to ensure that your science communication practices are inclusive and equitable?
- What kinds of institutional efforts would better support inclusive and equitable science communication practice?

CHAPTER 9

Self-Care and Collective Care

SCIENCE IN THE TWENTY-FIRST CENTURY is a notoriously stressful and competitive endeavor. If you are at a university, you might be teaching, writing grants, dealing with bureaucracy, and trying to find bits of time to do research. If you are at a government agency, you are almost certainly navigating politics while trying to do the best work possible. Graduate students and early-career researchers report immense mental health challenges related to everything from racism[1] to hostile campus environments to a job market that gets more competitive by the day.[2] Burnout is common.[3] The uncertainty related to the emergence of the coronavirus has only intensified many of these stressors,[4] particularly for minoritized researchers and practitioners.[5]

For science communicators in particular, pressures can come from any number of places, including working in emotional and contested environments and being inundated by what is generally bad news for much of the day while managing social media accounts that never sleep and a news cycle that doesn't end. There is online harassment of all forms—in the condescending comments sections of articles or unwanted sexual advances in direct messages from colleagues and strangers alike. There is also the potential for visible mistakes or misperceptions that can cost a job or even a whole career. These kinds of experiences can of course lead to burnout and health issues.

In addition, the many practitioners in these pages are living in the places they work and experiencing the same things as others in their communities. As Sarah Watson says in the opening chapter, "I have the same wet feet as my neighbors." This is true for me as well—when there's drought or wildfire in California, I am living through it while communicating about it. It is difficult to take time or space away from the things that are happening in your own backyard, and the consequences are profound. From climate change to food security to disaster, science communicators are engaged in systems change work on complex matters that will not be solved in a single career.

This is where self-care becomes critical. Self-care is generally considered to be the ways that we attend to our physical and emotional selves. Ideally it is restorative, not something that makes us feel worse or inadequate. Contemporary self-care is critiqued for having becoming too closely aligned with consumerism, perfectionism, and a host of other drawbacks,[6] and yet it is absolutely necessary for many science communication practitioners who are immersed in big efforts to address fundamental societal issues.

There are many ways to think about the interplay between the things that we do to care for ourselves and our communities and the work that we do to transform the systems that we are all a part of. As the experiences that follow will detail, however, there are strong themes that emerge in the types of practices that people undertake to stay healthy and engaged in their long-term work. These practices fall into several major categories that overlap: embodied practices like exercise and breath work, connecting with others, time for contemplation and reflection, establishing boundaries, pursuing hobbies, and space for joy and celebration.

Care for Self, Care for Others

"I try to get up early most mornings and accomplish something physically challenging, which makes it much easier to think about all the mentally challenging things I have to do later in the day," says Lydia Jennings. She is a doctoral candidate at the University of Arizona, where she studies soils and mining reclamation.

In addition to her pursuing her graduate education, Jennings is also an avid runner. An "Indigenous trail-running scientist" in her words, a triad she says has helped to both keep her sane and allow her to work on many of the subjects that she is concerned about in unique ways. "When my physical training is going well, I'm also able to work more efficiently."

Jennings spent a lot of time outside growing up, learning gardening skills from her dad and walking her dogs in the high mountain desert of northern New Mexico. She started running cross-country in high school and early on was inspired by the extensive history of running in Indigenous communities, including her own Pascua Yaqui (Yoeme) and Huichol (Wixáritari) nations. She has continued to run and is now able to connect back to Indigenous youth through trail running. Jennings has talked to young Indigenous runners about using running as a way to "get to know the earth and engage with science at the same time," helping students to see that running can be a way of making observations about one's surroundings, a first step in scientific practice. She also developed a trail-running series that supports Indigenous runners in getting more familiar with the many trails around Tucson.

The niche Jennings found emerged in part from her science communication efforts. On any given day, she can be found posting educational tweets and Instagram photos and stories filled with inspiring scenes of the desert. But it's not just the beauty that she captures. Jennings also puts a great deal of time and effort into explaining what she's seeing when she is out running. While she focuses on signs of mining in the western United States, she has an infectious desire to learn about every place she travels and can be seen posting on anything from traditional gathering of saguaro cactus fruit by the Tohono O'odham to tribal food sovereignty.

A recent trip to Alaska reinforced how much learning there is to be done simply by sharing land, food, and wisdom with other Indigenous people. There, Jennings was inspired by the co-created sense that everyone had expertise to bring to the table, whether it came from formal training or not. She describes the Indigenous researchers she was with in Alaska, trying to learn from their elders about how to bring back traditional moose-tanning practices. "They would ask elders direct questions about how to clean a moose hide, and the elders wouldn't answer. But then the researchers would bring an actual moose hide and try various techniques, only to then have elders jump in to correct the youth immediately. That kind of hands-on teaching has a lot to offer in the sciences, including respecting different ways of knowing, learning, and expertise," Jennings says.

Her hope is that sharing these kinds of experiences will inspire others to be and learn together on the land in new ways. "You figure out what is around you by knowing what to look for," she says. "You might be a lifelong runner and spend much time observing the land you're running on but not realize what you are observing. Or you might know a lot about something like the history of mining in the southwestern United States but not know how to look for signs of mining in a particular place because you don't spend much time out on the land the way trail runners do. I'm trying to support people in seeing the landscape while they're enjoying it recreationally."

Jennings has put a great deal of thought and strategy into how she navigates communicating about science. She does not separate her personal interests from her professional and says, "I don't feel like I should have to. I do science and I run, one because of the other. If I try to separate them, it leads to feeling siloed, and I just can't operate like that. Why should the way I talk about science be separate from who I am?"

She says that this way, she gets to be both her "nerdy and outdoorsy" selves, and that's what works for her.

Jennings is also clear about why she chooses social media as a specific communication tool. "Academia is a place with a lot of privilege, and I feel obligated to share with my community and others in ways that are more equalizing. I love getting questions from community members or when people share information with me. I know some of the information they share is because I am an Indigenous scientist who understands their shared cultural values."

She notes that it takes a lot of uncompensated time and work to create visually appealing and informative social media stories. Jennings also has to put effort into staying safe online. "I never post my location until I'm back at home, and I have to think carefully about the kinds of information that I post. I also have to be careful with the identities of some friends and family who just don't want to be online at all."

As she's spent more time in trail-running communities, talking about mining and other controversial land uses, she finds herself caught in the middle. Jennings says a lot of the places that she runs—and hikes and skis—are old mining towns located throughout the western United States. These places may have complicated histories of Indigenous removal and extractive industries, and Jennings's identity and interests place her squarely in the midst of the tensions that come with the territory. Meanwhile, many of these communities are increasingly turning to outdoor recreational activities as ways to sustain themselves, and Jennings wants to be a part of that future.

"I love to be outside, I love learning more about the world every day, and I am also committed to being an active part of a more just future. For me, science is about service, particularly for marginalized communities. And that kind of service is an incredible responsibility with some thorny topics like land dispossession and data sovereignty that require a lot of patience and commitment."

Jennings sees herself doing science and running well into the future, in part inspired by mentors like Karletta Chief and by what she considers kind advice from advisors to make sure to protect herself and her time, particularly as a marginalized scholar. She was told early on to pick her battles, but it's still a lifelong struggle. "Some days, the microaggressions are so hard to deal with and I don't feel like I belong in science at all, and other days, the fear of missing out is so hard when I have to say no to things. But I'm learning to listen to myself both in running and in academia so that I can stay healthy and in it for the long term."

Self-Care in Practice

Self-care practices are diverse. Jennings has found ways to incorporate movement, be outside, and develop community simultaneously. Like Jennings, many people turn to embodied practices for sustenance. It hardly matters what the practice is as long as it helps you to connect with yourself. For some, like Jennings, it might be running. For others it might be surfing or dancing, and others might be drawn toward stillness and breath work.

Contemplative practices such as meditation are common. There is a sense that contemplative practices involve specific settings and activities—a cushion, an altar, absolute quiet. However, anything that can help to quiet the mind can be considered contemplative. For some people, that is, again, movement based. Maybe it's tai chi or gardening. For others it might be playing the guitar or birding. For me, at this point in my life, it's journaling, walking, and yoga. Regardless of the activity, many practitioners have found that their contemplative and reflective practices that allow them to access a meditative state of mind can also be helpful in professional settings. For example, the discernment that comes from meditation practice can help a person to become a better listener.

In addition, the ability to put bounds around work by having a set workday or a flexible schedule or saying no to avoid being overbooked is key. Finding time for joyful activities like vacation travel or play and celebrating when good things happen are antidotes to being overwhelmed. Sleep, time away from work, being able to call on others and be called upon in return are all fundamental self-care practices. They are also lifelong practices.

One of the most common themes to emerge when talking with science communication practitioners is the value of connection and community. Jennings described finding refuge in creating and sustaining community. Connecting with others, whether friends and family, mentors, or peers, is a lifeline. Communities outside of work can help ground and provide time away from work and a way to maintain identity aside from a career. Communities in a work context can help to make work more fulfilling. Sometimes there is value in integrating the two, as the experiences shared below explore.

The Connection Is the Point
"Sometimes I feel caught between two worlds," says Tiffany Carey. As the coordinator for habitat and education at the National Wildlife Federation in Detroit, Michigan, Carey has one foot in the environmental world and another in the spiritual community she was born into as a PK, or pastor's kid.

Born and raised in Detroit, Carey says she "didn't grow up hiking and camping in the same way a lot of my peers did; I just wanted to make my own community a better place. I could have done that in a lot of different ways—as a social worker, a teacher, a pastor, a lawyer—but I've ultimately felt driven to connect with people through green spaces that also serve community needs."

As a student at the University of Michigan, Carey landed in ecology through an undergraduate research opportunity program. "I came into science in an unusual way. I wasn't immersed in the outdoors or obsessed with space as a kid. In college, I got interested in this idea that you could come up with a question and answer it, and from there, I found my way to ecology." Carey went on to develop a community science project focused on overlooked city spaces: vacant lots. She was particularly drawn to understanding the role these green spaces played in producing allergens that directly affect the health of the people living near them.

That project led to a position as a Growing Green Initiative community organizer in Baltimore, Maryland, where Carey worked with several different groups, including residents and city agencies, to create a vision for these unused green spaces. "In that position, I learned a lot about dialogue—even things that seem simple, like the benefits of trees on city streets, can be fraught when they are just sort of dictated to people without their input—and how scientific information can be leveraged to create change that communities want to see."

Now back in Detroit, Carey is committed to staying and working in the place she loves. In her role with the National Wildlife Federation, she is working on a project called Sacred Grounds, which weaves together her faith and science backgrounds. The project is focused on partnering with faith communities to create wildlife habitat, link spiritual practices with environmental stewardship, and help integrate these practices throughout communities.

In this role, Carey says she has found something that helps to bridge her professional and spiritual lives. "This kind of place-based learning is what I love; it gives us a point of connection to be outside together, viewing it as spiritual practice. But it's a kind of science communication that is not that traditional way of scientists talking to a journalist and getting something in the paper about a traumatic story to get people to be moved to do something. It encompasses peace and justice work and the co-benefits that wildlife conservation brings, which include everything from healing to things like fresh air and places for community gatherings."

That kind of bridging work is complex. Carey says, "Actually bringing the people that I love and the communities that I've grown up with into understanding about why the environment is so important can be hard. I feel like if I can get the people closest to me to understand why I do the work I do, make that change in my own family system, that will mean that I've achieved what I'm setting out to do."

Carey is also still working to find her place in the environmental world. "I am a young Black woman who is frequently in a lot of primarily white spaces. I want to be there, and I love the work, but there are still a lot of structural barriers in environmental professions for people of color. I don't have the same networks, the same cultural reference points; it can make fitting in hard. Even when people are well intentioned, we've got a long way to go in making environmental organizations representative of the people they want to serve."

To stay in what can be consuming work, Carey has a suite of practices she relies on. "I have so many practices I use to stay grounded, stay connected with what I feel like I'm here to do. Therapy is valuable. I run. I journal, a lot, and that's something I learned from my dad, how to self-reflect. My friends, my family, my partner, traveling, all of that helps keep me grounded. But I also crave solitude, and I've had to learn how to have better boundaries and not overburden myself. Ultimately, though, it's healing internally to be able to work authentically and give to my community. Those things have to go together for me."

Doing Good, Being Well

Like Jennings and Carey, many people become scientists and science communicators because they want to be of service. And yet the training process itself—and the hidden curriculum that accompanies it—can make people forget why they followed that path to begin with. Sharon Dobie describes the fears she and her peers had entering their medical residencies, writing, "My friends and I feared that the stresses of training and the hidden curriculum would whittle away the wonderment and gratitude of our early years. We worried that empathy would become secondary to surviving the intense pressures of residency."[7] Later in her career, as she taught medical students herself, she saw that fear reflected back as students grapple with the same issues but found refuge in the same kind of relationship-centered work that has been described in this book.

Part of what Dobie finds useful is a full-spectrum take on training and professional development that encourages senior practitioners to model good practices for trainees, similar to the work Aradhna Tripati described in chapter 8. As Gail Silverstein, the law professor, said, "It's difficult to try to teach students to avoid burnout, to focus on well-being, when we aren't great at modeling it. We know that people pay attention to our actions more than our words, and so we're still not quite in alignment with what we're trying to teach when it comes to sustainable careers."

Burnout is common in many fields and certainly in those that might fall under the banner of "helping professions," like counseling, nursing, and social work. Although science is generally not considered a helping profession—as the science communication practitioner narratives throughout this book indicate—many practitioners are in fact acting in helping roles. This means that some of the challenges facing, for instance, therapists or international aid workers are worth reviewing.

Indeed, practitioners are dealing with circumstances that are emotional, contentious, and uncertain, with high stakes for society. In an individualist culture, taking on extra responsibility and playing the hero can feel tempting. Yet many of the issues science communicators face are systemic, meaning not under our individual control and not able to be solved individually. They must be addressed collectively.

Here, a paradox emerges: the work to address systemic issues is intense and demoralizing for individuals, and we have to take care of ourselves and others to be able to attend to our collective work. Neither succeeds without the other. Unfortunately, a false dichotomy is set up between what is considered to be individual effort and care and collective or structural reform efforts.

The term *self-care* has its roots in medical care and can itself be traced to an emphasis on individual versus societal, factors that contribute to health declines; prioritizing individual wellness over universal health care would be one example.[8] During the civil and women's rights movements, however, the term was reclaimed, and Audre Lorde is recognized for creating a touchstone for self-care, writing in the late 1980s that "caring for myself is not self-indulgence, it is self-preservation, and that is an act of political warfare."[9]

Prioritizing Wellness

Tashiana Osborne is a doctoral candidate at the Scripps Institution of Oceanography.[10] Her research is focused on what are known as atmospheric rivers, a term for systems that can result in large rain- and snowstorms or—if they don't arrive—droughts. "I love that my research has implications for public safety, water and food security, and the economy. Learning to be a better communicator is a constant when you work on topics that directly affect people and communities."

Osborne has already had some incredible career experiences, including working as a storm chaser, interning at NASA, and attending United Nations climate negotiations. In addition, she says it is imperative to her to empower younger students and share the possibilities of science fields with marginalized communities. "It's important to acknowledge that, starting from a young age, students from underrepresented groups can face additional, complex barriers stemming from inequities. These also exist in higher education programs, including STEM. Diverse perspectives contribute to better science, innovation, and communication, but diversity initiatives alone are not enough without including focus on justice, equity, inclusion," she says.

Her commitment to youth in STEM comes in part from her own experiences. At various times, Osborne has navigated violence, economic instability, and food hardship. "I'm learning, although it's difficult, that the situations I was born into don't necessarily determine my future or what I invest in for the present. I've lived in different places and experienced various types of chaos. At times, in the professional realm, I've been targeted, excluded, and treated as 'less than.' I've been one of only a few people of color in some schools, classes, organizations, and workplaces," she says. "Even though it's challenging to share these vulnerabilities, I want to convey some of these struggles in addition to the positive highlights I typically share so that younger people can see there are many paths to becoming a scientist."

Throughout her life, Osborne found solace in teachers and coaches who nurtured her interests. "I participated in many clubs and activities at school. I would become sad and miss my teachers when I had to be away too long," she says. "My single mother worked to enroll my brother and me in better public schools farther outside our neighborhoods. This isn't always an option for others. I battled serious health issues and difficulties at home, but I found supportive people along the way, both within and outside STEM, to whom I am grateful. I went to St. Cloud State University, which happened to be the only school in Minnesota that had a meteorology program. Because of the overall lasting encouragement I gained as an undergraduate student, it honestly ended up being one of the most unexpectedly 'best' decisions I've made."

Like many undergraduates, Osborne wasn't entirely sure what her major would entail; she'd been drawn to it because she loved to lie in the grass and look at clouds and their shapes, which she says was a way to escape.

> I was, however, intimidated by the math and science. I saw I could succeed with practice and patience but thought of myself as naturally more of a words person. I'd like to say it became easier, but it remained difficult. While managing coursework, I was working multiple campus jobs to support myself, fighting for

my health while also helping loved ones, and remaining involved in professional skill-building organizations. I was told by physicians to cut back, and on more than one occasion. School and related opportunities, however, provided me a sense of hope that I was not willing to lose at the time.

Some challenges increased for Osborne in graduate school:

It can be hard at times to be in academic environments where simply being yourself takes a certain level of boldness. I've faced many aggressions that I hesitate to call "micro" because they can carve deep wounds over time. I've had people question my worthiness, question my name, or comment on how I dress. Partly because I have been considered a minority throughout my experiences, it took me a long time to understand that as anti–woman of color, anti-Black, anti-'otherness' hostility, not something I did wrong. I put a lot of energy into trying to perfect myself and constantly appease others, only to realize I wasn't actually the problem.

Osborne says, "It's pain on top of pain because meanwhile I've seen brothers and sisters being targeted, used as bootstraps, killed in the streets or their homes over and over, and it hasn't stopped. I've had to realize that, whether multiracial or not, we really don't have equal protections and equal rights. And my science degrees won't protect me. I've needed to build a certain level of resilience within the academic realm while also putting effort into calling for accountability and change, speaking up, and paying it forward."

Above all, Osborne says, she's learned to prioritize her health and wellness. "I have to set healthy boundaries. I can't pull all-nighters or overly stress before every meeting. It's hard on my body and can affect emotional and mental health too. I continue building tools and skills and setting realistic expectations to help embrace and adjust to ever-changing priorities. I have to make my best attempt at managing stress to be well enough to keep up with responsibilities, pursue goals, and help lift up others along the way."

Like Jennings and Carey, Osborne says that one thing that helps is connecting with others. "I call on champions I value and admire, even if they aren't in my field or our work doesn't exactly align, who I respect and can learn from. They are people I relate to in unique, even profound, ways as multidimensional, heart-centered individuals. I see the ways they do great work in the world, fight for themselves and others, and still show up as human."

In addition, Osborne reminds herself of the "strength and sacrifices of my ancestors and what they might think of the world today. I'm also building on a sense of passion and curiosity shared with my brother and mom and a sense of security and structure from my aunt, uncle, grandmother, and teachers. I talk with my grandma a lot. She's not at a point where she's climbing this ladder and can remind me that when I'm not well, I should rest. She recognizes this as a long-term journey, not a sprint. And if I'm trying to sprint for too long, it just will lead to burnout."

Osborne says her long-term goal is to explore questions that directly address problems society faces and share in ways that help contribute to solutions. "That way," she says, "there's a chance those individual puzzle pieces can fit into broader efforts focused on protecting people and ecosystems and, optimistically, improving the world we live in. But I also realize I have to stay well enough to do that work to begin with."

Psychological Support

Physical health issues are not the only potential problems facing science communication practitioners. Mental health can also be challenging. Although there is not much data on science communicators in particular, in the sciences and in academia, mental health issues are well documented, especially for graduate students.[11] I am not a mental health professional, and I encourage everyone who needs it to seek care from licensed providers. Formal psychotherapy has been a significant part of my own healing that began after my oral exams in my doctoral training program and continues today.

I have found that my science communication practice focused on emotional and contentious topics adds an extra layer to the normal stresses of life and work. Indeed, practitioners can face mental health challenges related to their subject matter. Priya Shukla, the UC Davis graduate student in chapter 2, spoke with journalist David Corn of *Mother Jones* about her work on climate change. Shukla said that a confluence of events led her to realize that she was "'emotionally exhausted' by the toll of constantly scrutinizing the 'huge tragedy' happening in the oceans. 'I did not want to experience that fatigue,' she says, 'because then I wouldn't want to do this work anymore.' She decided to see a therapist."[12]

However, after seeing a therapist once, Shukla decided therapy was not for her and has found other ways to cope with her work, including a weekly practice of a day of self-care. "I'm beginning to realize that taking care of your own mental health is beneficial to the people around you—by taking care of yourself, the people around you are better able to take care of you too (because needing support is okay!)," she tweeted recently.[13]

As Shukla's experience indicates, formal therapy is not for everyone—time, cost, culture, the process itself can all be reasons for people to seek other kinds of support. Luckily, there are formal and informal supporting practices and networks that may benefit individuals for whom psychotherapy is not a good fit.

Inner Practice and Outer Work

"This pandemic has made it a hard time to be a Black scientist," says Mila Marshall, the doctoral candidate at the University of Illinois at Chicago from chapter 3.

> I've spent a lot of time recently working on water shutoffs, particularly in Black communities. Cook County has the largest population of African Americans of any county in the nation. There has been so much disproportionate impact and harm done to the Black community here. Racism has created risk in so many different ways.
>
> We're living in polluted industrial corridors and dealing with respiratory illnesses and asthma, living in food deserts and dealing with obesity and high blood pressure. There are many people here that are essential workers and that has put them at even bigger risk from the coronavirus.

Marshall adds, "Then there are people who can't access the resources they need to keep themselves healthy and safe, including water. So the pandemic has given people an opportunity to see the intersections involved in being Black and poor, being Black and waterless."

Marshall says that one of her most valuable places of respite during a particularly difficult time has been her yoga practice and community:

> I'm so happy that yoga found me. It has given me tangible practices related to ethical living, meditation, and concentration. I live by the principle of not causing harm. It sounds simple, but it's not. Science has caused harm in Black communities. Even the kinds of questions that don't get asked cause harm, so I think a lot about how to be a scientist and not cause harm.
>
> With help from my yoga practice, I realized I had allowed academia to shape me into a version of myself I didn't want to be. Part of my work is to reflect on how the world has decided a Black body should be seen and heard and then to realize I don't have to follow that pattern.

Marshall says, "I'm very aware of how my gender and race and age and being a single mom impacts how people treat me and see me, and it also impacts how I treat myself and how I relate to other people as well. It's a whole scene."

Learning to work with her emotions has been valuable. "So often our emotions rise because of differing perceptions, so it's always helpful to take a minute to make sure you're seeing what you're seeing and feeling what you're feeling," says Marshall.

> My therapist told me if you're uncertain, try not to choose the most hurtful, negative interpretation immediately. That kind of practice is helpful to me personally and allows me to have more complex conversations with people about science, about separating feelings from their effects. It's not for everyone, but I decided as a Black woman who does know how to facilitate and finesse and support and hold space, I can communicate with a sense of compassion and awareness of people's capacity. You get to decide how heavy a topic is for you, how deep you want to go into it.

Marshall says that what keeps her going despite how heavy things can be is her commitment to the world. "I enjoy waking up every day knowing that my career honors all living things. There is equity in the balance of nature. No one thing is better than the other, all things are interconnected. You can't touch one without touching others. It's a ripple effect," she says.

"I wouldn't want to wake up and try and sell somebody a product that was created in a way that not only causes harm to the environment but oppresses people. I'm fortunate that I realized I'm not in control here and that there's a way to enjoy having a human experience without causing so much damage and destruction," Marshall says. "It's odd how we spend our lives trying to feel valuable when the rest of nature is just, like, the value is in being alive. The value is breathing."

Establishing Boundaries

While connecting with others is invaluable, having enough time for oneself—whether in work or play—is also crucial for many. Saying no can become a necessity, whether it's by blocking off time every day or week for specific activities, deciding that certain kinds of travel are not worth their effort, or determining that an extra talk given without an honorarium is not an option. It can also cause anxiety, as Jennings said: even though she understands she needs to say no sometimes, it's not always easy, particularly early in one's career or when you are simply excited about many different things. It can also have real consequences, so understanding that it is valuable, but needs to be counterbalanced by the costs of doing whatever the thing is, which can turn out to be higher than we imagine.

Taking breaks—whether to exercise or read for pleasure or nap—is another common way to place boundaries around work. It might be a walk around the block after finishing a piece of work or taking the time to make a healthy meal or do nothing at all. Taking breaks from screens has also become a crucial, if not extremely difficult, task for many science communicators.

Another way of thinking about putting boundaries around work as an act of self-care is to let go of control and perfectionism. The psychologist Donald Winnicott is well known for developing the idea of the "good enough parent"[14]—a concept that came in response to what many perceived as parenting advice that focused on "goodness" or perfectionism. This perspective can be applied to much of life—you can be the "good enough science communicator." That can feel particularly difficult in high-stakes environments, and yet not allowing things to be "good enough" can lead to controlling behavior and burnout.

Similarly, operating from a position of nonurgency is valuable. Of course, there are times when things must be urgent, but most of us are not truly working in first responder roles—and even first responders require breaks to continue to be effective.

Saying no to events and engagements is one thing, but being transparent about the work that you will do, how much you need to get paid or otherwise rewarded for it so that it feels well compensated, is another. In many ways, it's a "say no to say yes" approach. Although it's relevant at every career stage, discretion of this type frees us to progress in a career by saying no to some kinds of work so that we can say yes to others.

Walking the Talk
"At the end of my career, I would like to be able to say I supported small landowners in their ongoing implementation of the sustainable forest management practices that keep them and their families, as well as the land, healthy for generations," says Dan Stark, an assistant professor of practice with Oregon State University in Forestry and Natural Resources Extension. "What I've learned over time is that my technical training is only one piece of being able to do that work effectively."

Stark's career is focused on forestry education, mainly with small woodland owners. He supports their efforts to comply with regulations, learn new practices, and incorporate climate change into management efforts. While that might sound straightforward enough, Stark describes a multilayered and ongoing process of both professional and personal development that helps him to stay engaged in what can be delicate work.

"Even though my job is about providing information, it's also about a lot of other things. It's about me as a person in this community and the relationships I have with the other people here. They have to know they can trust me before much else can happen," Stark says. "I am relatively new here but am lucky because our institution is generally trusted and well liked, even though sometimes people see us as a bit too aligned with industry. Trust is complicated."

To gain and maintain trust, Stark says he's approached this job with lessons he's learned along the way. "There are several contentious, polarized issues I'm working on," he says. "Earlier in my career, I might just have jumped in and tried to get the environmental and forestry sectors in the same room. But I've learned I need to take time to understand the community first. So I'm focusing on listening and relationship building first."

At the same time, Stark notes, he has to navigate being himself in a place that he hopes to live and work for the long term. "Every time I go somewhere new, I have to go through this process of coming out again. When I first came to Oregon, I thought, I'm not going to talk about being gay, I'm not going to talk about my politics. But it was too hard to suppress my own humanity. I love to talk, I can be loud, and that's me. I'm mindful about how I express myself, but I also have started to understand that relationships are mutual—I can't completely give up myself to keep others comfortable. It's an ongoing negotiation and learning process."

Stark says that a big part of what keeps him going is being in the outdoors. "At big turning points in my life, I'll just go sit for entire days on the beach and not see another soul. I sit there with my dog and watch the waves roll in and process stuff," he says. "I'll see a family of eagles fly by, and that connection to the planet provides me with a lot of space and strength that I can't get in any other way. I work on forest management all day, but being outside gives me a chance to touch back into why I do the work I do to begin with.

"I've also found some professional development opportunities helpful," he says. "Classes on Crucial Conversations and facilitation have taught me instrumental skills like language I can use for speaking more effectively and how to listen well. Even with that training, you have to walk that talk. The actual practice is an area of growth at all times."

Stark says it's also been helpful to find a community of people who are dealing with similar things:

> There are more and more people talking about the skills that we need in addition to technical training. I've been fortunate to have supportive colleagues. The support from women in

particular has been so helpful. At many of the key points in my life that have led to big changes, there's been a woman at the center of that, and I just bow down because they do so much while navigating so many challenges.

I also recently watched a group of Black birders and naturalists talking about Black Birders Week and how to be an antiracist during a virtual Audubon event. The gathering was inspiring, and it made me reflect on the fact that I've learned to listen well but don't always know how to take the next step and act.

Stark concludes by saying:

> The speakers emphasized that you have to commit to antiracist work and work on yourself so you are in it for the long haul. That resonated with me.
>
> I see that kind of commitment as important because I've had to go through that process in my own work. To do it well, I have had to commit to figuring myself out. I am a white man, and I'm not trying to say my experience is at all equivalent to what women, women of color, people of color go through. There is a way that admitting, even to myself, how much I struggled has allowed me to have greater compassion for other's struggles and to join together to alleviate suffering wherever we can.

Taking Care

Self-care, too, tends to get wrapped up in yet another to-do list of things that are supposed to perfect us: mindfulness, dutiful exercise, treatments that ensure perpetual youth. However, self-care as described here is about taking care of yourself as you collaborate with others on the kinds of systemic issues that invite a collective approach. Self-care is dynamic—what you need will change throughout your life and career and may not look anything like it does for others—and so self-compassion is also key. Sometimes it's rest and sleep, other times it's getting physically exhausted by joyful movement, sometimes a nourishing meal with friends, sometimes actually finishing that one last task so you don't have to worry about it anymore.

Remember that self-care is not about developing the ability to remain in toxic environments indefinitely. Indeed, it is crucial for institutions to support environments that ease the need for self-care as a remedy. All the meditation in the world cannot make up for unsustainable institutional environments. Science communicators need to be supported by their institutions as they do their work in emotional and contentious environments.

Ultimately, self-care can be collective care.

Takeaways
- Science communication practitioner careers can be challenging, owing to factors like working within a highly charged media environment on emotional and contentious topics with high stakes.
- Self-care—taking care of one's physical and emotional self—is vital as science communicators engage with what are often deeply systemic issues that are unlikely to be "solved" during their careers.
- Common self-care practices include embodied or movement-based practices, connecting with others, time for contemplation and reflection, establishing boundaries, pursuing hobbies and other-than-work interests, and space for joy and celebration.
- Seek mental health support from licensed professionals when needed.
- It is important that self-care activities do not become another chore, another way of feeling like we're not perfect enough or doing enough, but instead are truly regenerative.

Questions for Exploration
- What issues did reading this chapter raise for you?
- What kinds of supports do you have in your work and personal life?
- What kinds of self-care routines do you have in place?
- What kinds of self-care routines could you create?
- How can you work in solidarity with others to ensure everyone can take care of themselves?

CHAPTER 10

What More Is Possible?

THE MANY PRACTITIONER NARRATIVES IN THIS BOOK demonstrate that the on-the-ground experience of science communication practitioners is complex and takes place on shifting terrain. The issues science communicators are contending with are emotional and many are contentious, with high stakes for practitioners and communities alike. Working in charged contexts is even more complex because of who individual practitioners are and the positions that they hold. We began the twenty-first century with a focus on "getting the message out"—often from elite scientists to elite journalists and decision-makers—and two decades in, it is apparent that strategy alone will not suffice. Instead, practitioners are invited to get off the stage and wade into the "mess" along with everyone else. It was simply an illusion that many weren't there already.

If we drop the pretense of science communication as an intellectually armored endeavor carried out by elites and instead commit to relating with others using some of the tools described here—listening with accountability, working with conflict, and understanding trauma—we get to ask a more interesting question: What more is possible? Hopefully that question, alongside the many experiences you've read about here and faced in your own lives, sparks your imagination in ways that

I can't yet imagine. In the meantime, I offer some closing thoughts on training and practice, prioritizing reward and protection over incentives, ethics and accountability, equity and inclusivity, and supporting relational work.

Training and Practice
"In medicine, it would be unthinkable to let a doctor with no clinical training treat patients. The stakes are too high. The same should be true for scientists engaging with communities," says Yanna Lambrinidou, the anthropologist we've heard from in these pages. She notes that the time spent in scientific training should be coupled with training in community engagement that teaches the limitations of technical expertise, the technical and moral relevance of community knowledge, and the value of oversight and accountability mechanisms for preventing abuses of scientific power.

This need for rigorous training offers a deep invitation to develop graduate programs that advance science communication *in practice*. It is not necessary to be overly prescriptive in describing exactly what those programs would look like, as the particulars will be up to many people with diverse considerations. Perhaps it is a full-scale professional doctoral degree in science or research communication; maybe it is something more akin to a dual degree program or emphasis. In any case, there are three key elements that such programs should include: (1) meaningful interaction with science communication practitioners and programs, (2) training in relational science communication, and (3) treating work with communities outside academia with respect and accountability.

Indeed, the key to any practice-based program is regular exposure to practitioners and practice-based work. This could include professors of practice who maintain professional positions that allow them to continue to develop their own skill sets while working with students. It should include substantive time spent in practice-based settings with dedicated support from experienced mentors. These kinds of programs will take work and funding but have multiple benefits. They would allow organizations exposure to different types of science communicators than they might normally consider hiring. In addition, students need opportunities to develop relationships with mentors outside academia and to adjust to the culture of nonacademic organizations to help them secure positions postgraduation. These opportunities need to be equitably available for all interested students.

Second, the relational approaches described here can provide some framing for what a training program might begin to encompass. Some courses that would be necessary for such a curriculum already exist on many campuses—the history of science, feminist studies, ethnographic methods—and could be woven together in interdisciplinary practitioner training programs. However, practical science communication topics, including many of the ones we've already explored, are not necessarily going to emerge from existing programs and will need to be developed, and preferably in collaboration with or taught by practitioners.

Finally, as Lambrinidou stresses, working with communities outside academia must be recognized as an exercise in power that requires the same standards of intellectual rigor and professional competence as any other application of scientific expertise. Care is needed because when there is no training in community engagement and no mechanism for preventing improper use of scientific authority, science communication practitioners may inadvertently cause more harm than good. This point cannot be overstated.

It is also worth noting that much current science communication training comes in the form of informal short courses and workshops, and Sunshine Menezes from chapter 8 says that in those settings, "we need to make sure we are in fact hearing from and valuing a much more representative set of people. We need to shift who is doing the training, why it's being done, and what it covers. There's always going to be some training to help scientists connect with journalists, but we need to be doing much more than that at this point."

Beyond training, as practitioners move into early- and mid-career positions as science communicators, their concerns evolve. Because science communicators frequently work outside academia, or in varied positions within it, they lack professional support other than from informal groups of colleagues. For the field of science communication to advance, it must move beyond the focus on training students and allow for discussion of more advanced practices. A professional society or semi-regular conference focused on science communication is one possibility; others include new initiatives and networks aimed at bridging the divide between academics and practitioners within existing societies.

In addition, in many fields, continuing education is standard. While it can turn into a bureaucratic exercise, continuing education is an important way of engaging with new ideas and new people, as well as a more informal way of advancing work. Overall, an intergenerational perspective on training and supporting science communication practitioners would benefit everyone.

From Incentives to Protection and Reward
The issues facing society are controversial, they are emotional, and engaging with them is risky for science communicators. The traditional ways of communicating science are often helpless in the face of this, and the longer it takes to realize it, the more time is wasted. Not having the right tools can also lead practitioners to avoid controversial work completely or leave that work to practitioners from groups who are most affected. Equity and inclusion work, for instance, regularly falls to marginalized groups.

Among the most ubiquitous, repetitive lines in papers and presentations on science communication is that it "must be incentivized" for individuals. I would like to propose that the word *incentivized* be dropped and replaced with "protected and rewarded." It is clear from the focus on incentives that the people eager to do this work are, bizarrely, not the focus of much advice on the topic. Many of us have never needed incentives—we do it because we are good at it, we enjoy it, we think it's valuable, it's an imperative because of who we are—and the incentives conversation simply invisibilizes that. At some point, it might be valuable to understand how so singular a focus on incentives for scientists who don't or won't communicate and approximately zero regard for protecting those who do originated, but that is for another day.

The narratives throughout this book demonstrate that many science communication practitioners are in fact in less protected classes in terms of employment and many other factors. That means practitioners are doing this work at great personal and professional risk, often in the face of what sociologist Tressie McMillan Cottom describes as a mismatch between institutional and cultural power.[1] Because many practitioners are not and will not be in tenure-protected positions, they need other forms of protection.

That might come in the form of institutions acknowledging this complex communications environment, understanding who is actually vulnerable in their organizations, and ensuring that those people are protected when they are exposed to significant risks. Beyond formal institutional protections, legal defense such as the Climate Science Legal Defense Fund, which provides "support and resources to scientists who are threatened, harassed or attacked for doing their jobs," offers a model that could be extended.[2]

Science Communication Ethics

What might an ethic like "first, do no harm" look like in science communication and engagement? Moving forward, professional ethics for science communication and engagement practitioners would be valuable to discuss,[3] although of course professional ethics frameworks have their own limitations. Lawyers and doctors have professional ethics; it does not mean they are sufficient or cover the range of ethical dilemmas, but they do set a baseline.

Research ethics are simply not adequate to cover practice-based dilemmas. The Institutional Review Board process addresses some subset of concerns about human subjects in research settings. But when human subjects protocols are taken out of the research context and into community-engaged work, there is little relevant to draw from. For example, when working with communities, what is a benefit for one community may create a material loss for another. Similarly, gaining the trust of one community may come at the expense of losing the confidence of another—from what I have seen, there is no universal trustworthiness, nor should there be.

As Lambrinidou pointed out, it is also important that scientists and communicators working with communities resist the temptation to view themselves as "heroes," "martyrs," or "saviors." Because many people go into this work wanting to be helpful, they may have trouble recognizing when their well-intentioned involvement has veered into causing harm. This is why ethical guidelines, motivated by justice, or at the very least transparent discussions about ethics, are necessary.

Equitable, Inclusive, and Just Science Communication

No matter the language, which is evolving by the day, centering equity, inclusivity, and justice in science communication work is, at the least, the kind of bridge that Menezes described in chapter 8 to a better science communication future. There is no doubt that race, gender, sexuality, age, location, and many other factors pertaining to both science communicators and the people they work with influence science communication practice and outcomes. The brilliant work of practitioners in chapter 8 represents the future of science communication.

It is therefore important that while working on these controversial subjects, which practitioners are highly invested in and care deeply about, they must also have resources to take care of themselves and avoid, or have time to recover from, burnout. This does not mean that the burden of addressing systemic issues is on individuals to learn to cope with or endure, rather that institutions must support the work and workers central to their mission.

Supporting Relationship-Centered Work
As is the case with the inclusivity and equity work described above, it is also notoriously difficult to get support for the time and effort that it takes to carry out relational work. Amy Cardinal Christianson, the fire scientist from chapter 7, says that while connecting with First Nations is fundamental in her work, and the agency she represents is supportive, it is difficult to get funding for the relationship building that her projects require.

Similarly, Julian Reyes, American Association for the Advancement of Science (AAAS) Science and Technology Policy Fellow, notes a similar trend; while agencies are supportive of working with communities, that support rarely comes in the form of financial backing. When talking with anybody doing relational work, you will eventually hear a lament about the years needed to do the work well versus the resources available to do it.

While funding support is welcome, given the uncertainty in science, academia, and beyond, it is also important to point out that there are ways to conduct preliminary relational work that do not require much monetary investment. For example, setting relational goals for even short meetings to normalize the value of relationship building can begin to shift practice. Beginning standard presentations with relational interactions—invitations to breathe together, to introduce each other, to welcome one another—can help to normalize these kinds of practices.

There are also many more advanced approaches to relational science communication that can be explored and put into action. With the support of two small funders, I have held intimate workshops with diverse groups of practitioners from various fields who have different groundings in relational work. The conversations, ideas, and relationships themselves have been rich and full of potential and have led to further collaborations.[4]

Again: What More Is Possible?
This book has focused on practitioner experiences, and while those are undoubtedly valuable, it would likewise be of great benefit to better understand the experiences of the people whom science communicators are interacting with. Although some science communicators are interrogating the effects of their work, a wide-ranging look at who is and is not served by current science communication efforts would be revealing.

I hope the personal narratives in this book have demonstrated the need to update the model of what science communication is and can be. There is a vital and exciting world of opportunity waiting to be explored in the move away from performance-based science communication and toward community-accountable, relational engagement.

Notes

Introduction

1. Faith Kearns, "The Stepping Stones of Integrating Emotions into Practicing Science," *On Being* (blog), March 14, 2015, https://onbeing.org/blog/the-stepping-stones-of-integrating-emotions-into-practicing-science/.

Chapter 1. Science Communication from the Ground Up

1. Eric S. Blake et al., "Tropical Cyclone Report Hurricane Sandy," National Hurricane Center, February 12, 2013, https://www.nhc.noaa.gov/data/tcr/AL182012_Sandy.pdf.
2. Lindy Orthia, "Strategies for Including Communication of Non-Western and Indigenous Knowledges in Science Communication Histories," *Journal of Science Communication* 19, no. 2 (2020): 1–17, https://doi.org/10.22323/2.19020202.
3. Hewitt, "Science Communications," *Science* 114, no. 2953 (1951): 134–35, https://doi.org/10.1126/science.114.2953.134.
4. Rose Hayden-Smith and Rachel Surls, "A Century of Science and Service," *California Agriculture* 68, no. 1 (January 2014): 8–15, https://doi.org/10.3733/ca.v068n01p8.
5. See, for example, Robert Lee and Tristan Ahtone, "Land-Grab Universities," *High Country News*, March 30, 2020, https://www.hcn.org/issues/52.4/indigenous-affairs-education-land-grab-universities.
6. See, for example, Carmen V. Harris, "'The Extension Service Is Not an Integration Agency': The Idea of Race in the Cooperative Extension Service," *Agricultural History* 82, no. 2 (2008): 193–219, https://doi.org/10.3098/ah.2008.82.2.193.

7. See, for example, Teresa Carr, "Revisiting the Role of the Science Journalist," Undark, July 15, 2019, https://undark.org/2019/07/15/science-journalism-communications/.
8. See, for example, Molly J. Simis et al., "The Lure of Rationality: Why Does the Deficit Model Persist in Science Communication?," *Public Understanding of Science* 25, no. 4 (2016): 400–414, https://doi.org/10.1177/0963662516629749.
9. Orthia, "Strategies for Including Communication."
10. See, for example, John Schwartz, "A Climate Change Evangelist," *New York Times*, October 11, 2016.
11. See, for example, Shahzeen Z. Attari, David H. Krantz, and Elke U. Weber, "Statements about Climate Researchers' Carbon Footprints Affect Their Credibility and the Impact of Their Advice," *Climatic Change* 138, nos. 1–2 (2016): 325–38, https://doi.org/10.1007/s10584-016-1713-2.
12. See, for example, Citizens Climate Lobby (website), https://citizensclimatelobby.org/.
13. See, for example, "National Network for Ocean and Climate Change Interpretation," Climate Interpreter, https://climateinterpreter.org/about/projects/NNOCCI.
14. See, for example, Yale Program on Climate Change Communication (website), https://climatecommunication.yale.edu.
15. See, for example, Renee Lertzman, "How Can We Talk about Global Warming?," *Sierra*, July 19, 2017, https://www.sierraclub.org/sierra/how-can-we-talk-about-global-warming.
16. See, for example, Edward Maibach, Matthew Nisbet, and Melinda Weathers, *Conveying the Human Implications of Climate Change: A Climate Change Communication Primer for Public Health Professionals* (Fairfax, VA: George Mason University Center for Climate Change Communication, 2011).
17. See, for example, Mary Annaise Heglar, "I Work in the Environmental Movement. I Don't Care If You Recycle," *Vox*, May 28, 2019, https://www.vox.com/the-highlight/2019/5/28/18629833/climate-change-2019-green-new-deal; and Frances Roberts-Gregory, "My Petrochemical Love," Alliance for Affordable Energy, April 23, 2020, https://www.all4energy.org/the-watchdog/my-petro-chemical-love.
18. See, for example, "COMPASS Message Box," COMPASS, https://www.compassscicomm.org/leadership-development/the-message-box/.

19. Jessica Eise, "What Institutions Can Do to Improve Science Communication," *Nature*, December 13, 2019, https://doi.org/10.1038/d41586-019-03869-7.
20. See, for example, Erving Goffman, *The Presentation of Self in Everyday Life* (New York: Anchor Books, 1959), 17.
21. Jason A. Delborne, "Constructing Audiences in Scientific Controversy," *Social Epistemology* 25, no. 1 (2011): 67–95, https://doi.org/10.1080/02691728.2010.534565.
22. "Diamond Holloman: Women in Science Wednesdays," Endeavors, May 23, 2018, https://endeavors.unc.edu/diamond-holloman/.
23. "Diamond Holloman."
24. Eve Elliot, "How Diamond Holloman Finds Resilience in Lumberton," Carolina Center for Public Service, August 9, 2019, https://ccps.unc.edu/how-diamond-holloman-finds-resilience-in-lumberton/.
25. Max Liboiron, "Exchanging," in *Transmissions: Critical Tactics for Making and Communicating Research*, ed. Kat Jungnickel (Cambridge, MA: MIT Press, 2020), 1–15.
26. See, for example, Judy L. Meyer et al., "Above the Din but in the Fray: Environmental Scientists as Effective Advocates," *Frontiers in Ecology and the Environment* 8, no. 6 (2010): 299–305, https://doi.org/10.1890/090143; and Mary H. O'Brien, "Being a Scientist Means Taking Sides," *Bioscience* 43, no. 10 (1993): 706–8.
27. "Our History," Nature Conservancy, accessed June 23, 2020, https://www.nature.org/en-us/about-us/who-we-are/our-history/.
28. "History," Union of Concerned Scientists, accessed June 23, 2020, https://www.ucsusa.org/about/history.
29. "About PEER—Protecting Public Employees," PEER, June 12, 2020, https://www.peer.org/about-us.
30. Marvel (@DrKateMarvel), "On one hand, being a vocal climate science communicator can be detrimental to your academic job prospects," Twitter, March 20, 2019, 9:43 a.m., https://twitter.com/DrKateMarvel/status/1108363472096710657.
31. Myhre, "Climate Scientists, Mourning Earth's Losses, Should Make Their Voices Heard," *Guardian*, May 19, 2016, https://www.theguardian.com/environment/climate-consensus-97-per-cent/2016/may/19/climate-scientists-mourning-earths-losses-should-make-their-voices-heard.

32. Kahanamoku, "The Fight for Mauna Kea and the Future of Science," Massive Science, July 17, 2019, https://massivesci.com/notes/mauna-kea-thirty-meter-telescope-colonialism-astronomy/.
33. Peter Weingart and Lars Guenther, "Science Communication and the Issue of Trust," *Journal of Science Communication* 15, no. 5 (2016): 1–11, https://doi.org/10.22323/2.15050301.

Chapter 2. Science Communication Careers Today

1. Stephanie E. Hampton and Stephanie G. Labou, "Careers in Ecology: A Fine-Scale Investigation of National Data from the U.S. Survey of Doctorate Recipients," *Ecosphere* 8, no. 12 (2017): 1–11, https://doi.org/10.1002/ecs2.2031.
2. National Science Board, "SEH Doctorates in the Workforce, 1993–2013," National Science Foundation, 2019, https://nsf.gov/nsb/sei/infographic2/#nsb-statement.
3. Blickley et al., "Graduate Student's Guide to Necessary Skills for Nonacademic Conservation Careers," *Conservation Biology* 27, no. 1 (2012): 24–34, https://doi.org/10.1111/j.1523-1739.2012.01956.x.
4. Carolyn A. F. Enquist et al., "Foundations of Translational Ecology," *Frontiers in Ecology and the Environment* 15, no. 10 (2017): 541–50, https://doi.org/10.1002/fee.1733.
5. See, for example, Morgan Meyer, "The Rise of the Knowledge Broker," *Science Communication* 32, no. 1 (2010): 118–27, https://doi.org/10.1177/1075547009359797.
6. See, for example, Alex Ketchum, "Report on the State of Resources Provided to Support Scholars against Harassment, Trolling, and Doxxing while Doing Public Media Work and How University Media Relations Offices/Newsrooms Can Provide Better Support," Public Scholarship and Media Work, July 14, 2020, https://publicscholarshipandmediawork.blogspot.com/p/report.html.
7. To be clear, here I am addressing issues of research-based information that is perceived as controversial, not protection of hate speech. See, for example, Farhana Sultana, "The False Equivalence of Academic Freedom and Free Speech," *ACME: An International Journal for Critical Geographies* 17, no. 2 (2020): 228–57, https://www.acme-journal.org/index.php/acme/article/view/1715.

8. See, for example, Huo Jingnan, "Why There Are So Many Different Guidelines for Face Masks for the Public," NPR, April 10, 2020, https://www.npr.org/sections/goatsandsoda/2020/04/10/829890635/why-there-so-many-different-guidelines-for-face-masks-for-the-public.
9. Michael Specter, "How Anthony Fauci Became America's Doctor," *New Yorker*, April 20, 2020, https://www.newyorker.com/magazine/2020/04/20/how-anthony-fauci-became-americas-doctor.
10. These kinds of tensions have long existed within academia and in many ways are to be expected. See, for example, la paperson, *A Third University Is Possible* (Minneapolis: University of Minnesota Press, 2019), https://doi.org/10.5749/9781452958460, which describes, among many other things, the dual nature of the university captured in the opening line: "Within the colonizing university also exists a decolonizing education."
11. Brad Plumer and Coral Davenport, "Science under Attack: How Trump Is Sidelining Researchers and Their Work," *New York Times*, December 28, 2019, https://www.nytimes.com/2019/12/28/climate/trump-administration-war-on-science.html.
12. H. Holden Thorp, "Stick to Science," *Science* 367, no. 6474 (January 10, 2020): 125.
13. McMillan Cottom, "Risk and Ethics in Public Scholarship," *Inside Higher Ed*, 2011, https://www.insidehighered.com/blog/8948.
14. McMillan Cottom, "'Who Do You Think You Are?': When Marginality Meets Academic Microcelebrity," *Ada: A Journal of Gender, New Media, and Technology*, no. 7 (2015), https://doi.org/doi:10.7264/N3319T5T.
15. McMillan Cottom, "Risk and Ethics in Public Scholarship," *Inside Higher Ed*, 2011, https://www.insidehighered.com/blog/8948.
16. Zevallos, "Sociology of Public Harassment Prevention Policies," Other Sociologist, 2017, https://othersociologist.com/sociology-public-harassment-prevention-policies/.
17. Ibid.

Chapter 3. Navigating Facts and Feelings in Science Communication

1. Donna Haraway, "Situated Knowledges: The Science Question in Feminism and the Privilege of Partial Perspective," *Feminist Studies* 14, no. 3 (1988): 575–99, https://doi.org/10.2307/3178066.

2. Sheila Jasanoff, *States of Knowledge: The Co-production of Science and Social Order* (London: Taylor & Francis, 2004), 93–94.
3. Bruno Latour, "Scientific Objects and Legal Objectivity," in *Law, Anthropology and the Constitution of the Social: Making Persons and Things*, ed. Alain Pottage and Martha Mondy, trans. Alain Pottage (Cambridge: Cambridge University Press, 2004), 73–113.
4. See, for example, Lewis Raven Wallace, *The View from Somewhere: Undoing the Myth of Journalistic Objectivity* (Chicago: University of Chicago Press, 2019); and Wesley Lowery, "A Reckoning over Objectivity, Led by Black Journalists," *New York Times*, June 23, 2020, https://www.nytimes.com/2020/06/23/opinion/objectivity-black-journalists-coronavirus.html.
5. Martinez, "The View from Nowhere," October 15, 2019, in *The View from Somewhere*, produced by Ramona Martinez, podcast, MP3 audio, 19:00, https://www.lewispants.com.
6. Lesley M. Head and Theresa Harada, "Keeping the Heart a Long Way from the Brain: The Emotional Labour of Climate Scientists," *Emotion, Space and Society* 24 (2017): 34–41.
7. Irus Braverman, *Coral Whisperers: Scientists on the Brink* (Oakland: University of California Press, 2018), 154.
8. Kahanamoku, "Fight for Mauna Kea" (see chap. 1, n. 32).
9. Oreskes, "The Scientist as Sentinel," *Limn*, no. 3 (June 2013), https://limn.it/articles/the-scientist-as-sentinel/.
10. Corn, "It's the End of the World as They Know It," *Mother Jones*, July 8, 2019, https://www.motherjones.com/environment/2019/07/weight-of-the-world-climate-change-scientist-grief/.
11. Liz Bondi, "Making Connections and Thinking through Emotions: Between Geography and Psychotherapy," *Transactions of the Institute of British Geographers* 30, no. 4 (2005): 433–48.
12. Sultana, "Suffering for Water, Suffering from Water: Emotional Geographies of Resource Access, Control and Conflict," *Geoforum* 42, no. 2 (2011): 163–72, https://doi.org/10.1016/j.geoforum.2010.12.002.
13. Tosca (@climategal84), "Aware that the material that we covered in class can be demanding and anxiety-inducing, I recognize the importance of ending the semester on a note of optimism," Twitter, November 20, 2019, 5:53 p.m., https://twitter.com/climategal84/status/1197286910873481217.

14. Hoffman, "Why Transitioning a Farm from One Generation to the Next Is Trickier Than Ever," *Counter*, March 25, 2019, https://thecounter.org/family-farm-transitions-land-transfer-extension-agents/.
15. Wolfe, "Flush Your Disgust. We Can't Let Emotions Dampen Our Water Policies," *Globe and Mail*, March 22, 2019, https://www.theglobeandmail.com/opinion/article-flush-your-disgust-we-cant-let-emotions-dampen-our-water-policies/.
16. Parsafar, "Exploring the Relationship between Emotions and Water Issues," interview by Faith Kearns, *Confluence*, August 22, 2018, https://ucanr.edu/blogs/blogcore/postdetail.cfm?postnum=28017.
17. Ahmed, *The Cultural Politics of Emotion* (New York: Routledge, 2010), 3.
18. Adapted from Faith Kearns, "Water Challenges for California's Small Farm Community," *Confluence*, September 11, 2017, https://ucanr.edu/blogs/blogcore/postdetail.cfm?postnum=25150.
19. Cullis-Suzuki, "The UN Experience," *Sea around Us*, no. 59 (May/June 2010), http://www.seaaroundus.org/wp-content/uploads/2010/01/SAUP59.pdf.

Chapter 4. Relating

1. Portions of this chapter are adapted from Faith Kearns, "A Relational Approach to Climate Change: Working with People and Conflict," in *Climate Change across the Curriculum*, ed. Eric J. Fretz (Lanham, MD: Lexington Books, 2015), 219–34.
2. Jasanoff, "Testing Time for Climate Science," *Science* 328, no. 5979 (June 2010): 695–96, https://doi.org/10.1126/science.1189420.
3. Jasanoff, 695–96.
4. Oddly, in fields like ecology, while relationships between organisms and their environments are central, interpersonal relationships have for the most part remained unaddressed.
5. See, for example, Mary Catherine Beach and Thomas Inui, "Relationship-Centered Care: A Constructive Reframing," *Journal of General Internal Medicine* 21, no. S1 (2006), https://doi.org/10.1111/j.1525-1497.2006.00302.x.
6. See, for example, Susan L. Brooks and Robert G. Madden, "Relationship-Centered Lawyering: Social Science Theory for Transforming Legal Practice," *University of Puerto Rico Law Review* 78 (2009), https://ssrn.com/abstract=1193009.

7. See, for example, Oddbjrn Leirvik, *Interreligious Studies: A Relational Approach to Religious Activism and the Study of Religion* (London: Bloomsbury, 2015), 5.
8. See, for example, Bondi, "Making Connections," 433–48 (see chap. 3 n. 11).
9. See, for example, Jessica Benjamin, "A Relational Psychoanalysis Perspective on the Necessity of Acknowledging Failure in Order to Restore the Facilitating and Containing Features of the Intersubjective Relationship (the Shared Third)," *International Journal of Psychoanalysis* 90, no. 3 (2009): 441–50, https://doi.org/10.1111/j.1745-8315.2009.00163.x; and Philip M. Bromberg, "Shrinking the Tsunami," *Contemporary Psychoanalysis* 44, no. 3 (2008): 329–50, https://doi.org/10.1080/00107530.2008.10745961.
10. See, for example, Carol Gilligan, *In a Different Voice: Psychological Theory and Women's Development* (Cambridge, MA: Harvard University Press, 1982), xiv.
11. See, for example, Stephen A. Mitchell, *Relational Concepts in Psychoanalysis: An Integration* (Cambridge, MA: Harvard University Press, 1995), 141.
12. See, for example, Mark Epstein, *Going on Being: Buddhism and the Way of Change* (New York: Broadway Books, 2002), 32.
13. See, for example, bell hooks, *Feminist Theory: From Margin to Center* (Cambridge, MA: South End, 2000), 93.
14. See, for example, Sandra Harding, *The Feminist Standpoint Theory Reader: Intellectual and Political Controversies* (New York: Routledge, 2003), 26.
15. Kimmerer, *Braiding Sweetgrass: Indigenous Wisdom, Scientific Knowledge, and the Teachings of Plants* (Minneapolis: Milkweed Editions, 2013), 57–58.
16. Simpson, "Land as Pedagogy: Nishnaabeg Intelligence and Rebellious Transformation," *Decolonization: Indigeneity, Education & Society* 3, no. 3 (2014): 1–25.
17. TallBear, "Caretaking Relations, Not American Dreaming," *Kalfou* 6, no. 1 (2019), https://doi.org/10.15367/kf.v6i1.228.
18. Todd, "Relationships," Theorizing the Contemporary, *Fieldsights*, January 21, 2016, https://culanth.org/fieldsights/relationships.
19. Yazzie, "Water Is Life, Relationality, and Tribal Sovereignty," interview by Faith Kearns, *Confluence*, October 24, 2017, https://ucanr.edu/blogs/blogcore/postdetail.cfm?postnum=25499; see also Melanie K. Yazzie,

"Water at the Center in a Time of Tension and Possibility," interview by Faith Kearns, *Confluence*, November 6, 2017, https://ucanr.edu/blogs/blogcore/postdetail.cfm?postnum=25598.

20. Kimberly M. S. Cartier, "Keeping Indigenous Science Knowledge Out of a Colonial Mold," *Eos*, December 11, 2019, https://eos.org/articles/keeping-indigenous-science-knowledge-out-of-a-colonial-mold.
21. David-Chavez and Michael C. Gavin, "A Global Assessment of Indigenous Community Engagement in Climate Research," *Environmental Research Letters* 13, no. 12 (2018): 123005, https://doi.org/10.1088/1748-9326/aaf300.
22. David-Chavez and Gavin, 123005.
23. Ermine, "The Ethical Space of Engagement," *Indigenous Law Journal* 6, no. 1 (2007): 193–203, https://jps.library.utoronto.ca/index.php/ilj/article/view/27669/20400.
24. Allan N. Schore, "The Right Brain Implicit Self: A Central Mechanism of the Psychotherapy Change Process," in *Knowing, Not-Knowing & Sort-of-Knowing: Psychoanalysis and the Experience of Uncertainty*, ed. Jean Petrucelli (London: Routledge, 2010), 177–202.
25. See, for example, George Marshall, "Understand Faulty Thinking to Tackle Climate Change," *New Scientist*, August 13, 2014, https://www.newscientist.com/article/mg22329820.200-understand-faulty-thinking-to-tackle-climate-change.html; and Dan Kahan, "Fixing the Communications Failure," *Nature* 463, no. 7279 (2010): 296–97, https://doi.org/10.1038/463296a.
26. Renee Lertzman, "In Climate Change, Psychology Gets Lost in Translation," *Pacific Standard*, June 14, 2017, https://psmag.com/environment/in-climate-change-psychology-often-gets-lost-in-translation.
27. Susanne C. Moser and Lisa Dilling, "Communicating Climate Change: Closing the Science-Action Gap," *Oxford Handbooks Online*, 2011, https://doi.org/10.1093/oxfordhb/9780199566600.003.0011.
28. Faith Kearns, "From Science Communication to Relationship-Building: Contemplative Practice and Community Engagement in the Environmental Sciences," *Journal of Environmental Studies and Sciences* 2, no. 3 (2012): 275–77, https://doi.org/10.1007/s13412-012-0083-y.
29. Peter T. Coleman, "Conflict, Complexity, and Change: A Meta-framework for Addressing Protracted, Intractable Conflicts—III," *Peace and Conflict:*

Journal of Peace Psychology 12, no. 4 (2006): 325–48, https://doi.org/10.1207/s15327949pac1204_3.

30. Kearns, "From Science Communication," 275–77.
31. Rosemary Randall, "Loss and Climate Change: The Cost of Parallel Narratives," *Ecopsychology* 1, no. 3 (2009): 118–29, https://doi.org/10.1089/eco.2009.0034.
32. Renee Lertzman, "Breaking the Climate Fear Taboo," *Sightline Institute*, March 12, 2014, http://daily.sightline.org/2014/03/12/breaking-the-climate-fear-taboo/.
33. See, for example, Enid Balint, "The Possibilities of Patient-Centered Medicine," *Journal of the Royal College of General Practitioners* 17, no. 82 (1969): 269–76; as well as the work the Black Panthers were doing to establish community health-care clinics around the same time period: Mary T. Bassett, "Beyond Berets: The Black Panthers as Health Activists," *American Journal of Public Health* 106, no. 10 (2016): 1741–43, https://doi.org/10.2105/ajph.2016.303412.
34. Balint, "Possibilities of Patient-Centered," 269–76.
35. Beach and Inui, "Relationship-Centered Care."
36. Dobie, "Viewpoint: Reflections on a Well-Traveled Path: Self-Awareness, Mindful Practice, and Relationship-Centered Care as Foundations for Medical Education," *Academic Medicine* 82, no. 4 (2007): 422–27, https://doi.org/10.1097/01.acm.0000259374.52323.62.
37. Dennis P. Stolle, David B. Wexler, and Bruce J. Winick, *Practicing Therapeutic Jurisprudence: Law as a Helping Profession* (Durham, NC: Carolina Academic Press, 2000), 88.
38. Brooks and Madden, "Relationship-Centered Lawyering."
39. Darshan Brach, "A Logic for the Magic of Mindful Negotiation," *Negotiation Journal* 24, no. 1 (2008): 25–44, https://doi.org/10.1111/j.1571-9979.2007.00165.x.
40. Davenport, *Emotional Resiliency in the Era of Climate Change: A Clinician's Guide* (London: Jessica Kingsley, 2017).
41. See, for example, Jon Kabat-Zinn, *Full Catastrophe Living: Using the Wisdom of Your Body and Mind to Face Stress, Pain, and Illness*, rev. ed. (New York: Bantam Books, 2013), 22–31.
42. One approach is described in Kearns, "Relational Approach," 219–34 (see chap. 4, n. 1), which provides more detail on expanding relational abilities

in terms of concrete relational processes, capacities, practices, and exercises that can be used in the classroom and beyond.
43. Ibid.
44. See, for example, Farhana Sultana, "Reflexivity," in *The International Encyclopedia of Geography: People, the Earth, Environment, and Technology*, ed. Douglas Richardson et al. (Hoboken, NJ: Wiley-Blackwell, 2017), 1–5.
45. Gupta, Campbell, and Cole-Weiss, "Cooperative Extension Can Better Frame Its Value by Emphasizing Policy Relationships," *California Agriculture* 73, no. 1 (2019): 11–18, https://doi.org/10.3733/ca.2018a0040.
46. Dobie, "Viewpoint: Reflections," 422–27.

Chapter 5. Listening

1. Lee, "2019 SACNAS Featured Speaker: Dr. Danielle N. Lee," YouTube video, 10:52, December 17, 2019, https://youtu.be/AC-g3jbjbNk.
2. Ibid.
3. Brooks, "Listening to Communities: A Bottom-Up Approach to Water Planning in California," interview by Faith Kearns, *Confluence*, May 7, 2018, https://ucanr.edu/blogs/blogcore/postdetail.cfm?postnum=27131.
4. Disadvantaged Communities Involvement Program, *Santa Ana River Watershed Community Water Experiences: An Ethnographic Strengths and Needs Assessment* (Riverside, CA: Santa Ana Watershed Project Authority, 2019), https://sawpa.org/wp-content/uploads/2019/12/OWOW-DCI-CDE-Santa-Ana-River-Watershed-Community-Water-Experiences_-An-Ethnographic-Strengths-and-Needs-Assessment-full-report-wit.pdf.
5. Zenger and Folkman, "What Great Listeners Actually Do," *Harvard Business Review*, July 14, 2016, https://hbr.org/2016/07/what-great-listeners-actually-do.
6. See, for example, Harry Weger et al., "The Relative Effectiveness of Active Listening in Initial Interactions," *International Journal of Listening* 28, no. 1 (February 2014): 13–31, https://doi.org/10.1080/10904018.2013.813234.
7. Kimmerer, *Braiding Sweetgrass*, 48 (see chap. 4, n. 15).
8. See, for example, Stephen Rollnick and William R. Miller, "What Is Motivational Interviewing?," *Behavioural and Cognitive Psychotherapy* 23, no. 4 (1995): 325–34, https://doi.org/10.1017/s135246580001643x.

9. Marshall B. Rosenberg, *Nonviolent Communication: A Language of Life* (Encinitas, CA: PuddleDancer, 2015), 6–8.
10. "Quick Statistics about Hearing," National Institute of Deafness and Other Communication Disorders, October 5, 2018, https://www.nidcd.nih.gov/health/statistics/quick-statistics-hearing.
11. Adapted from Faith Kearns, "Making the Invisible Visible: Connecting Home Water Use & City Infrastructure through Participatory Design," *Confluence*, May 7, 2017, https://ucanr.edu/blogs/blogcore/postdetail.cfm?postnum=24061.
12. See, for example, John D. Brewer, "The Ethnographic Critique of Ethnography: Sectarianism in the RUC," *Sociology* 28, no. 1 (1994): 231–44, https://doi.org/10.1177/0038038594028001014; and Carolina Alonso Bejarano et al., *Decolonizing Ethnography: Undocumented Immigrants and New Directions in Social Science* (Durham, NC: Duke University Press, 2019), 1–4.
13. Wutich, "Water Insecurity: An Agenda for Research and Call to Action for Human Biology," *American Journal of Human Biology* 32, no. 1 (July 2019), https://doi.org/10.1002/ajhb.23345.
14. Driftless Writing Center, "Stories from the Flood," 2019, https://www.wisconsinfloodstories.org/publications.
15. Tuck and Yang, "Decolonization Is Not a Metaphor," *Decolonization: Indigeneity, Education & Society* 1, no. 1 (2012): 1–40, https://jps.library.utoronto.ca/index.php/des/article/view/18630.
16. See, for example, Lambrinidou, "When Technical Experts Set Out to 'Do Good': Deficit-Based Constructions of 'the Public' and the Moral Imperative for New Visions of Engagement," *Michigan Journal of Sustainability* 6, no. 1 (2018), http://dx.doi.org/10.3998/mjs.12333712.0006.102.

Chapter 6. Working with Conflict

1. Adapted from Faith Kearns, "The Art and Science of Waiting," Hippo Reads, 2015, http://read.hipporeads.com/the-art-and-science-of-waiting/.
2. Sweeny, Patrick J. Carroll, and James A. Shepperd, "Is Optimism Always Best?," *Current Directions in Psychological Science* 15, no. 6 (2006): 302–6, https://doi.org/10.1111/j.1467-8721.2006.00457.x.

3. Sweeny, "Waiting Well: Tips for Navigating Painful Uncertainty," *Social and Personality Psychology Compass* 6, no. 3 (2012): 258–69, https://doi.org/10.1111/j.1751-9004.2011.00423.x.
4. Frans B. M. de Waal, "The Antiquity of Empathy," *Science* 336, no. 6083 (2012): 874–76, https://doi.org/10.1126/science.1220999.
5. Guy Riddihough et al., "Human Conflict: Winning the Peace," *Science* 336, no. 6083 (2012): 818–19, https://doi.org/10.1126/science.336.6083.818.
6. See, for example, Steve M. Redpath et al., "Understanding and Managing Conservation Conflicts," *Trends in Ecology & Evolution* 28, no. 2 (2013): 100–109, https://doi.org/10.1016/j.tree.2012.08.021; and Jürgen Scheffran et al., "Climate Change and Violent Conflict," *Science* 336, no. 6083 (2012): 869–71, https://doi.org/10.1126/science.1221339.
7. Adapted from Faith Kearns, "Working with Wild Horses and Water Controversies in Rural California," *Confluence*, July 17, 2017, https://ucanr.edu/blogs/blogcore/postdetail.cfm?postnum=24690.
8. Coleman, "Conflict, Complexity, and Change," 325–48 (see chap. 4, n. 29).
9. Coleman, 325–48.
10. Rina Faletti, "Understanding California's Water Culture," interview by Faith Kearns, *Confluence*, March 12, 2018, https://ucanr.edu/blogs/blogcore/postdetail.cfm?postnum=26597.
11. Doug Parker and Faith Kearns, "California's Water Paradox: Why Enough Will Never Be Enough," *California Journal of Politics and Policy* 8, no. 3 (February 2016), https://doi.org/10.5070/p2cjpp8331698.
12. Barrientos, "Highlighting the Gap between Policy Design and Policy Implementation," interview by Kathryn Stein, *Confluence*, December 2, 2019, https://ucanr.edu/blogs/blogcore/postdetail.cfm?postnum=38870.
13. Sifuentez, "Son of Farmworkers Turned Professor Focuses on Water Issues in California's Central Valley," interview by Faith Kearns, *Confluence*, August 14, 2018, https://ucanr.edu/blogs/blogcore/postdetail.cfm?postnum=27968.
14. Manne, *Down Girl: The Logic of Misogyny* (New York: Oxford University Press, 2017), 134–35.
15. Schulman, *Conflict Is Not Abuse: Overstating Harm, Community Responsibility, and the Duty of Repair* (Vancouver: Arsenal Pulp, 2017), 17.

Chapter 7. Understanding Trauma

1. See, for example, the American Psychological Association's resources on trauma, "Trauma," accessed October 19, 2020, https://www.apa.org/topics/trauma/.
2. See, for example, Lillian Comas-Díaz, Gordon Nagayama Hall, and Helen A. Neville, "Racial Trauma: Theory, Research, and Healing; Introduction to the Special Issue," *American Psychologist* 74, no. 1 (2019): 1–5, https://doi.org/10.1037/amp0000442.
3. See, for example, National Academies of Sciences, Engineering, and Medicine, "Addressing Historical, Intergenerational, and Chronic Trauma: Impacts on Children, Families, and Communities," in *Achieving Behavioral Health Equity for Children, Families, and Communities: Proceedings of a Workshop*, ed. Clarissa E. Sanchez et al. (Washington, DC: National Academies Press, 2019), 37–42, https://www.ncbi.nlm.nih.gov/books/NBK540764/.
4. Center for Substance Abuse Treatment-Trauma-Informed Care in Behavioral Health Services, "Understanding the Impact of Trauma," *Treatment Improvement Protocol Series*, no. 57 (2014), https://www.ncbi.nlm.nih.gov/books/NBK207191/.
5. Stacey Clettenberg et al., "Traumatic Loss and Natural Disaster: A Case Study of a School-Based Response to Hurricanes Katrina and Rita," *School Psychology International* 32, no. 5 (2011): 553–66, https://doi.org/10.1177/0143034311402928.
6. Tools like psychological first aid have been developed for disaster situations. See Julie A. Uhernik and Marlene A. Husson, "Psychological First Aid: An Evidence Informed Approach for Acute Disaster Behavioral Health Response," in *Compelling Counseling Interventions: VISTAS*, ed. Garry Richard Walz, Jeanne C. Bleuer, and Richard K. Yep (Alexandria, VA: American Counseling Association, 2009), 271–80.
7. Hill, "When Climate Change Knocks on Your Door, What Will You Say?," *Medium* (blog), December 22, 2017, https://medium.com/@tmhill/when-climate-change-knocks-on-your-door-what-will-you-say-5bbdbb68f4ed.
8. Robin Abcarian, "California Journal: They Survived Six Hours in a Pool as a Wildfire Burned Their Neighborhood to the Ground," *Los Angeles Times*, October 12, 2017, https://www.latimes.com/local/abcarian/la-me-abcarian-sonoma-fire-20171012-htmlstory.html.

9. Hill, "When Climate Change Knocks."
10. Bill Van Niekerken, "Wine Country Fire of 1964: Eerie Similarities to This Week's Tragedy," *San Francisco Chronicle*, October 10, 2017, https://www.sfchronicle.com/chronicle_vault/article/Wine-Country-fire-of-1964-Eerie-similarities-to-12267643.php.
11. Kia-Keating, "Trauma and Resilience in California Disaster Response," adapted from interview by Faith Kearns, *Confluence*, April 23, 2018, https://ucanr.edu/blogs/blogcore/postdetail.cfm?postnum=27000.
12. Many fields are taking up trauma-informed approaches. See, for example, Resmiye Oral et al., "Adverse Childhood Experiences and Trauma Informed Care: The Future of Health Care," *Pediatric Research* 79, nos. 1–2 (2015): 227–33, https://doi.org/10.1038/pr.2015.197; and Jill Levenson, "Trauma-Informed Social Work Practice," *Social Work* 62, no. 2 (April 2017): 105–13, https://doi.org/10.1093/sw/swx001.
13. Hankins, "California Indigenous Perspectives on Water and Fire Management," interview by Faith Kearns, *Confluence*, March 26, 2018, https://ucanr.edu/blogs/blogcore/postdetail.cfm?postnum=26710.
14. Eriksen and Hankins, "The Retention, Revival, and Subjugation of Indigenous Fire Knowledge through Agency Fire Fighting in Eastern Australia and California," *Society & Natural Resources* 27, no. 12 (October 2014): 1288–1303, https://doi.org/10.1080/08941920.2014.918226.
15. Christianson, Tara K. McGee, and Whitefish Lake First Nation 459, "Wildfire Evacuation Experiences of Band Members of Whitefish Lake First Nation 459, Alberta, Canada," *Natural Hazards* 98, no. 1 (November 2019): 9–29, https://doi.org/10.1007/s11069-018-3556-9.
16. Christianson (@ChristiansonAmy), "The First Nations Wildfire Evacuation partnership is very excited to release this summary infographic of our research results from a 6 year research project," Twitter, July 12, 2018, 11:39 a.m., https://twitter.com/ChristiansonAmy/status/1017478676722565120.
17. Office for Victims of Crime, "The Vicarious Trauma Toolkit: Introduction," Office for Victims of Crime, accessed July 6, 2020, https://ovc.ojp.gov/program/vtt/introduction.
18. Yan, "When Science Reporting Takes an Emotional Toll," Open Notebook, October 10, 2017, https://www.theopennotebook.com/2017/10/10/when-science-reporting-takes-an-emotional-toll/.

Chapter 8. Equitable, Inclusive, and Just Science Communication

1. This is beginning to change. See, for example, a special issue of the *Journal of Science Communication* that includes the following papers: Tania Pérez-Bustos, "Questioning the Feminization in Science Communication," *Journal of Science Communication* 18, no. 4 (2019), https://doi.org/10.22323/2.18040304; and Elizabeth Rasekoala, "The Seeming Paradox of the Need for a Feminist Agenda for Science Communication and the Notion of Science Communication as a 'Ghetto' of Women's Over-Representation: Perspectives, Interrogations and Nuances from the Global South," *Journal of Science Communication* 18, no. 4 (2019), https://doi.org/10.22323/2.18040307. See, in addition, Thomas Hayden et al., eds., "Inclusive Science Communication in Theory and Practice," special issue, *Frontiers in Communication* 5 (2020), https://www.frontiersin.org/research-topics/9107/inclusive-science-communication-in-theory-and-practice#articles.
2. Melissa C. Márquez and Ana Maria Porras, "Science Communication in Multiple Languages Is Critical to Its Effectiveness," *Frontiers in Communication* 5 (2020), https://doi.org/10.3389/fcomm.2020.00031.
3. Ellie Bothwell et al., "Women in Science Are Battling Both Covid-19 and the Patriarchy," *Times Higher Education*, May 15, 2020, https://www.timeshighereducation.com/blog/women-science-are-battling-both-covid-19-and-patriarchy.
4. Jennifer Gerson, "A Doctor & Protest Supporter Explains How Racism Is a Health Issue," Bustle, June 12, 2020, https://www.bustle.com/p/talking-about-racism-must-be-a-part-of-health-care-according-to-a-doctor-22978442.
5. Canfield et al., "Science Communication Demands a Critical Approach That Centers Inclusion, Equity, and Intersectionality," *Frontiers in Communication* 5 (2020), https://doi.org/10.3389/fcomm.2020.00002.
6. Nelson, "Interview with Ecologist Dr Joanna Nelson about Her Climate-Action Work at the Land-Sea Edge," interview by Promita Chakraborty, *Medium* (blog), February 5, 2020, https://medium.com/@500womensci.sfbay/interview-with-ecologist-dr-joanna-nelson-about-her-climate-action-work-at-the-land-sea-edge-d1577988f72e.
7. Serrato Marks and Bayer, "Our Disabilities Have Made Us Better Scientists," *Scientific American*, July 10, 2019, https://blogs.scientificamerican.com/voices/our-disabilities-have-made-us-better-scientists/.

8. Sarah Maslin Nir, "How 2 Lives Collided in Central Park, Rattling the Nation," *New York Times*, June 14, 2020, https://www.nytimes.com/2020/06/14/nyregion/central-park-amy-cooper-christian-racism.html.

9. See, for example, Carolyn Finney, *Black Faces, White Space: Reimagining the Relationship of African Americans to the Great Outdoors* (Chapel Hill: University of North Carolina Press, 2014); and J. Drew Lanham, "Birding While Black," Literary Hub, September 22, 2016, https://lithub.com/birding-while-black/.

10. Katie Langin, "'I can't even enjoy this.' #BlackBirdersWeek Organizer Shares Her Struggles as a Black Scientist," *Science*, June 5, 2020, https://www.sciencemag.org/careers/2020/06/i-cant-even-enjoy-blackbirdersweek-organizer-shares-her-struggles-black-scientist.

11. David Zax, "#BLACKandSTEM: The Hashtag as Community," *Fast Company*, March 4, 2014, https://www.fastcompany.com/3027122/blackandstem-the-hashtag-as-community.

12. Jafri (@JameelaJafri), "Legit, I've been a hijabi birder for over 20 years and this is the first time I've met a Black hijabi birder. I am so grateful to #BlackBirdersWeek for elevating," Twitter, June 1, 2020, 8:36 p.m., https://twitter.com/JameelaJafri/status/1267661214731005954?s=20.

13. Mike Gaworecki, "Conservationists Find Opportunity and Community amidst Current Crises," Mongabay, June 10, 2020, https://news.mongabay.com/2020/06/audio-conservationists-find-opportunity-and-community-amidst-current-crises/.

14. Andrea Thompson, "Black Birders Call Out Racism, Say Nature Should Be for Everyone," *Scientific American*, June 5, 2020, https://www.scientificamerican.com/article/black-birders-call-out-racism-say-nature-should-be-for-everyone/?amp=true.

15. Allen (@Glossopetalon), "Today is the first day of #BlackBotanistsWeek and we're highlighting our #BlackRoots through our #BlackBotanical Legacy. An unsung #Botany101 hero is Antione," Twitter, July 6, 2020, 6:41 a.m. https://twitter.com/Glossopetalon/status/1280134861307535360?s=20.

16. Turner (@_saveelephants), "I'm excited to go to grad school but also nervous. The lack of diversity in STEM has led me to question my capabilities and my place in biology," Twitter, June 5, 2020, 6:36 a.m. https://twitter.com/_saveelephants/status/1268899462493229057?s=20.

17. See full conference committees for 2018 at "2018 Planning Committee," InclusiveSciComm, https://inclusivescicomm.org/2018-planning-committee/; and for 2019 at "2019 Planning Committee," Inclusive SciComm, https://inclusivescicomm.org/2018-symposium/2019planning-committee/.
18. Moore, "Can We Talk?," kendallmooredocfilms, https://www.kendallmooredocfilms.com/can-we-talk.
19. Valdez-Ward, "Documenting California Drought as an Undocumented Researcher," interview by Faith Kearns, *Confluence*, December 3, 2018, https://ucanr.edu/blogs/blogcore/postdetail.cfm?postnum=28814.
20. Valdez-Ward and Cat, "Reclaiming STEM: Addressing the Gap in Diverse and Inclusive Trainings," Sister, June 17, 2019, https://sisterstem.org/2019/06/17/reclaiming-stem/.
21. Adapted from Faith Kearns, "Retaining Diverse Environment and Water Scholars through Community," *Confluence*, June 15, 2020, https://ucanr.edu/blogs/blogcore/postdetail.cfm?postnum=42758.

Chapter 9. Self-Care and Collective Care

1. See, for example, Nidhi Subbaraman, "Grieving and Frustrated: Black Scientists Call Out Racism in the Wake of Police Killings," *Nature* 582, no. 7811 (2020): 155–56, https://doi.org/10.1038/d41586-020-01705-x.
2. See, for example, Katia Levecque et al., "Work Organization and Mental Health Problems in PhD Students," *Research Policy* 46, no. 4 (2017): 868–79, https://doi.org/10.1016/j.respol.2017.02.008.
3. See, for example, Miguel A. Padilla and Julia N. Thompson, "Burning Out Faculty at Doctoral Research Universities," *Stress and Health* 32, no. 5 (2015): 551–58, https://doi.org/10.1002/smi.2661.
4. Chris Woolston, "Junior Researchers Hit by Coronavirus-Triggered Hiring Freezes," *Nature* 582, no. 7812 (2020): 449–50, https://doi.org/10.1038/d41586-020-01656-3.
5. Jessica L. Malisch et al., "Opinion: In the Wake of COVID-19, Academia Needs New Solutions to Ensure Gender Equity," *Proceedings of the National Academy of Sciences* 117, no. 27 (2020): 202010636, https://doi.org/10.1073/pnas.2010636117.

6. See, for example, Charlotte Lieberman, "How Self-Care Became So Much Work," *Harvard Business Review*, August 10, 2018, https://hbr.org/2018/08/how-self-care-became-so-much-work.
7. Dobie, "Viewpoint: Reflections," 422–27 (see chap. 4, n. 36).
8. See, for example, Aisha Harris, "How 'Self-Care' Went from Radical to Frou-Frou to Radical Once Again," *Slate*, April 5, 2017, http://www.slate.com/articles/arts/culturebox/2017/04/the_history_of_self_care.html.
9. Lorde, *A Burst of Light: And Other Essays* (Mineola, NY: Ixia, 2017), 130.
10. Osborne, "Researching California's Extreme Weather, Storm-by-Storm," adapted in part from interview by Faith Kearns, *Confluence*, July 15, 2019, https://ucanr.edu/blogs/blogcore/postdetail.cfm?postnum=30814.
11. Teresa M. Evans et al., "Evidence for a Mental Health Crisis in Graduate Education," *Nature Biotechnology* 36, no. 3 (2018): 282–84, https://doi.org/10.1038/nbt.4089.
12. Corn, "What Happens When You Can See Disaster Unfolding, and Nobody Listens?," *Mother Jones*, July 8, 2019, https://www.motherjones.com/environment/2019/07/weight-of-the-world-climate-change-scientist-grief/.
13. Shukla (@priyology), "I'm beginning to realize that taking care of your own mental health is beneficial to the people around you—by taking care of yourself, the people around you," Twitter, November 12, 2019, 11:00 a.m., https://twitter.com/priyology/status/1194329031916703744?s=20.
14. Winnicott, *The Child, the Family, and the Outside World* (Hammondsworth, UK: Penguin, 1971), 10.

Chapter 10: What More Is Possible?
1. McMillan Cottom, "'Who Do You Think You Are?'" (see chap. 2, n. 14).
2. See Climate Science Legal Defense Fund (website), https://www.csldf.org.
3. Science communication ethics is becoming a more widely discussed topic; see, for example, Fabien Medvecky and Joan Leach, *An Ethics of Science Communication* (Cham, Switzerland: Palgrave Pivot, 2019).
4. Faith Kearns, "Relationship First: An Experiential Workshop on Relationship-Centered Approaches to Climate Communication," *Science Unicorn* (blog), January 18, 2018, https://scienceunicorn.blogspot.com/2018/01/relationship-first-experiential.html.

About the Author

Faith Kearns is a scientist and science communication practitioner who focuses primarily on water, wildfire, and climate change in the western United States. Her work has been published in *New Republic*, *On Being*, *Bay Nature*, and more. She has been working in the science communication field for more than twenty-five years, starting with the Ecological Society of America and going on to serve as an AAAS Science and Policy Fellow at the US Department of State, manage a wildfire research and outreach center at the University of California, Berkeley, and bridge science and policy advocacy efforts at the Pew Charitable Trusts. She currently works with the California Institute for Water Resources. Kearns holds an undergraduate environmental science degree from Northern Arizona University and a doctorate in environmental science, policy, and management from the University of California, Berkeley.

Island Press | Board of Directors

Rob Griffen
(Chair)
Managing Director,
Hillbrook Capital

Deborah Wiley
(Secretary and Treasurer)
Chair, Wiley Foundation, Inc.

Decker Anstrom
Board of Directors,
Discovery Communications

David Bland
Chair,
Travois, Inc.

Terry Gamble Boyer
Author

Margot Ernst

Alison Greenberg

Marsha Maytum
Principal,
Leddy Maytum Stacy Architects

David Miller
President,
Island Press

Pamela Murphy

Alison Sant
Cofounder and Partner,
Studio for Urban Projects

Caelan Hidalgo Schick
Master of Public and International
   Affairs, 2021,
University of Pittsburgh

Ron Sims
Former Deputy Secretary,
US Department of Housing
   and Urban Development

Sandra E. Taylor
CEO,
Sustainable Business
   International LLC

Anthony A. Williams
CEO and Executive Director,
Federal City Council

Sally Yozell
Senior Fellow and Director
   of Environmental Security,
Stimson Center